Knowledge Quest

Planet Earth
and the Universe

The stars, dust and gas that make up a galaxy are held together by gravity. Turn to page 42 to find out more.

Knowledge Quest

Planet
Earth
and the Universe

Published by
The Reader's Digest Association Limited
London • New York • Sydney • Montreal

Contents

How to use Knowledge Quest

Knowledge Quest: Planet Earth and the Universe is a uniquely interactive reference book that brings you the essential facts and a wealth of wide-ranging information on the Earth, the Solar System and the Universe.

Knowledge Quest will make adding to your store of knowledge both interesting and fun. It builds into a highly illustrated reference series, with each volume delivering authoritative facts and many other significant things to know about a particular branch of knowledge.

Planet Earth and the Universe is packed with facts on space, our Solar System, and the Earth's physical geography and climate – all contained within the core reference section (pages 39 to 153). If you want to use the book as a straightforward source of information the contents page lists the major topics covered, while the detailed index (starting on page 161) allows you to look up specific subjects. If you simply like to browse, you will find that each fascinating piece of information leads you to discover another, and another, and another . . .

What's so special about Knowledge Quest?

The unique feature of **Knowledge Quest** is the set of quiz questions that can be used as an entertaining way to get into the reference information. You can use these to test out what you already know about the Earth and the Universe, and to lead you eagerly into finding out more.

How do the questions link to the reference section?

There are 100 quizzes of 10 questions each, which are graded and colour-coded for levels of difficulty (see right). Each question is accompanied by a page number, which is the page in the reference section where you will find both the answer and more information on that subject generally.

The answers to all questions relating to the topic of the spread – which will come from several quizzes – are listed by question number in the far-left column. A number (or sometimes a star) following each answer refers you to the box containing the most relevant additional information elsewhere on the spread. More than one box may be indicated, and sometimes none are especially relevant, in which case an additional detail is given with the answer.

The questions in each quiz lead you to several different pages and topics in the reference section, which is great for finding out more, but is not so convenient if you want to use the quizzes as straightforward quiz rounds. So, for all keen quizmasters, we also list the answers to all the questions in each quiz in the 'Quick answers' section which starts on page 154.

Each question in the quizzes at the front of the book is linked by a page number (given immediately below the question number) to a page in the reference section, where you will find the answer, plus additional information on the subject. In this example, question 160 is linked to page 120.

The answers to all the questions relevant to the overall topic of a reference spread are listed by question number in the left-hand column. Each answer is followed by a box reference showing where more information can be found. Sometimes, instead of a box reference, additional information is given with the answer.

Each box on the rest of the reference spread contains information about an aspect of the overall topic. In this example, the topic of pages 120–121 is 'Landslides and avalanches', and box ❹ features information on different types of snow avalanche. Sometimes, more than one box will be relevant to an answer.

From question . . . to answer . . . to discovering more

159
p.84
PREY SEEN on a mountain range between France and Spain.
Question

160
p.120
Mass of snow, ice and rocks sliding down a mountain gives you NAVAL ACHE.

120

QUESTION NUMBER

The numbers or star following the answers refer to information boxes on the right.

ANSWERS **Box reference**

160 Avalanche ❶ ❹

220 *Avalanche* (starring Hudson, 1978)

234 Landslide

An

Snow avalanches ❹

Most avalanches occur on slopes of 30– . Snow breaks free from the underlying sur and crashes downslope at speeds of 200 (120mph). There are two major types.

More information

POWDER Loose surface snow slides downslope a fan shape. Usually ally destruct

Colour-coding for quizzes of different levels of difficulty

A thousand questions are provided in a hundred themed quizzes. There are three levels of difficulty – **Warm Up**, **In Your Stride** and **Challenge**:

 Warm Up quizzes feature easy questions to get you into the swing. Children may enjoy these quizzes as much as adults.

 In Your Stride questions are pitched at a more difficult level than 'Warm Up' quizzes, requiring a little more knowledge of the subject.

 Challenge quizzes are the most difficult, often requiring in-depth or specialist knowledge. But have a go anyway – you never know what information may be tucked away in the recesses of your brain!

Two other categories, **All Comers** and **Multiple Choice**, include mixed-level questions, ranging from 'warm-up' level to 'challenge', and generally becoming harder as a round progresses.

 All Comers quizzes cover a range of levels of questions, from easy to hard. Everyone should be able to join in and see how they get on.

 Multiple Choice quizzes offer four possible answers to each question, only one of which is correct. They feature mixed-ability questions, generally arranged to become harder as a quiz progresses.

Special features

Star answers
One answer on each two-page spread is marked with a **star** in the answer column. This indicates a subject of special or unusual interest within the spread topic. (Occasionally, more than one answer in the answers column refers the reader to the star answer box.)

 Galileo

★ 122

Galileo
The theory of an **Earth-centred Solar System** was put forward by the Greek astronomer Ptolemy. It was first challenged by Nicolaus

Tie-breaker questions
Some spreads, but not all, feature tie-breaker questions:

Tie-breaker

These questions can be used by quizmasters in a quiz in the event of a tie, but they also contain information that expands on the answer for the interested reader.

Keeping score

The most straightforward **scoring method** is simply to award one point – or if you prefer, two points – for every correct answer.

If you are using the quizzes as rounds in a competition, you may find it easiest to look up the answers in the **Quick Answers** section at the back of the book.

For readers who would like to use this book as an information source for setting quizzes, a blank **Question sheet** and **Answer sheet** are provided at the back of the book. Quizmasters can photocopy the question sheet for their own use, and the answer sheet for distribution among contestants.

Colours of the world

In this opening quiz, all the answers refer to a colour.

1 p.136 Which planet in our solar system is known as the 'blue planet'?

2 p.62 Blackboards are traditionally made from plates of which fine-grained metamorphic rock?

3 p.128 Which popular cake is named after a forested region of south-west Germany?

4 p.84 Which one of the following states of America is called the Green Mountain State – Vermont, California or Alaska?

5 p.74 What colour is the mineral azurite?

6 p.126 Which is home to the white bear – the Arctic, Antarctic or Amazon?

7 p.44 Where are little green men supposed to come from?

8 p.76 What coloured fever were gold-prospectors in Australia said to have?

9 p.150 Which colourful organisation campaigns for the conservation of the environment?

10 p.76 Which white pigment once used in oil paints contains compounds of a soft metal?

Wet, wet, wet

Ten teasers on watery features and phenomena.

11 p.122 Originally a Japanese word, what is the term for a giant wave caused by an underwater earthquake or landslide?

12 p.130 In which layer of the atmosphere is rain produced – the thermosphere, troposphere or stratosphere?

13 p.108 Name the warm ocean current that flows from the Gulf of Mexico to north-west Europe.

14 p.112 Starting at the surface, put these ocean depth zones in order of descent: abyssal, hadal, twilight, sunlit.

15 p.94 Niagara Falls lie on the border between which two countries?

16 p.146 Name the visual phenomenon, a bit similar to a rainbow, that is caused by sunlight falling on fog?

17 p.92 In Greenland, the Petermann, Jungersen, Quarayaq and Frederickshab are all examples of what type of natural feature?

18 p.96 In which direction does the Nile flow: north to south, or south to north?

19 p.88 Which geographical feature are Superior, Disappointment and Surprise examples of?

20 p.108 How long is the shoreline of the Sargasso Sea?

Islands in the sun

The following ten questions are all about islands.

21 p.102 Which group of North Pacific islands is the 50th state of the USA?

22 p.102 Which group of islands includes Ibiza, Majorca and Minorca?

23 p.104 Which classic children's novel of a family wrecked on a desert island was updated in the TV series *Lost in Space*?

24 p.102 In which river would you find the Ile de la Cité?

25 p.104 Which island was named on December 25, 1643, by William Mynors?

26 p.104 The Aegean Islands are part of which country?

27 p.102 Pitcairn island was settled in 1790 by mutineers from which ship?

28 p.100 What is the literal meaning of the word 'peninsula'?

29 p.104 Which island in New York Bay was once an immigration centre?

30 p.102 Which legendary island was said to have been swallowed up by the Atlantic Ocean?

Animal planet

A series of questions on creatures around the world.

31 p.104 Tailless Manx cats are native to which island?

32 p.48 Laska and Berry were the first of their kind in space. What sort of creature were they?

33 p.76 The Lone Ranger's horse was named after which precious metal?

34 p.144 What term links a weather condition with Tintin's four-legged friend?

35 p.126 Which powerful double-coated dogs are used to pull sledges in the Arctic?

36 p.132 Weathervanes are often in the form of which bird?

37 p.90 In what type of environment would you normally find a dromedary?

38 p.80 Where were linnets or budgerigars used to detect gas?

39 p.102 Which songbird is named after a group of islands off the north-west coast of Africa?

40 p.70 The coelacanth was known only from fossils until one was found alive in 1938. What sort of creature is it?

For answers and more facts go to the page given below each question number.
For quick answers to complete quizzes 0 to 6 go to page 154.

Which continent?

A list of the seven continents is given below to help you.

41 p.82 Which continent contains two-thirds of the world's fresh water.

42 p.82 Which continent has the largest landmass?

43 p.82 Which continent has Death Valley as its lowest point?

44 p.90 Which continent contains the Sahara, Karoo, Kalihari and Namib deserts.

45 p.82 The name of which continent derives from ancient Greek meaning 'opposite the Bear'?

46 p.82 Which continent includes the Great Barrier Reef?

47 p.128 Which continent has vast treeless plains called pampas?

48 p.82 Which continent has the longest coastline?

49 p.128 Which continent contains the world's largest expanse of taiga?

50 p.82 Which is the flattest continent?

Africa
Antarctica
Asia
Europe
North America
Oceania (includes Australia)
South America

Movie magic

Ten questions with a film connection.

51 p.50 On which planet was the original movie *Planet of the Apes* set?

52 p.56 In the 1998 film *Deep Impact*, what is the Earth threatened by?

53 p.116 *Krakatoa, East of Java* features which two natural disasters?

54 p.50 In which alien invasion film did Jack Nicholson play the president of the USA?

55 p.134 In Jan de Bont's film *Twister*, what are the lead characters chasing?

56 p.76 What treasure is Humphrey Bogart looking for in *The Treasure of the Sierra Madre*?

57 p.94 What does Jeremy Irons fall over at the end of Roland Joffe's *The Mission*?

58 p.40 In the 1994 film *IQ*, which scientist did Walther Matthau play?

59 p.42 Which 'Quest' starring Tim Allen, was a movie spoof of a popular TV sci-fi series?

60 p.84 In which mountain range is the film *Cliff Hanger* set?

Pot luck

Select the correct option from the four possible answers.

61 p.74 The white powder that mixes with water to make casts for broken limbs is called Plaster of what?

A London	B Bonn
C Belfast	D Paris

62 p.142 Cirrocumulus clouds are also known as what type of sky?

A Kipper	B Mackerel
C Anchovy	D Smoked salmon

63 p.122 The rivers Ganges, Brahmaputra and Meghra merge on a giant flood plain in which of these countries?

A Bangladesh	B Venezuela
C Siberia	D Iraq

64 p.46 Huitzilopochti and Phoebus Apollo were both gods of what?

A The Sun	B The Moon
C The sea	D The wind

65 p.72 What do you get when two crustal plates pull apart?

A Rift	B Thunder
C Mountain	D Volcano

66 p.136 How many litres of water does an average washing-machine use for one wash?

A 75	B 7.5
C 0.75	D 750

67 p.42 The time the Sun takes to orbit the Galaxy is called a what?

A Cosmic year	B Calendar Year
C Leap Year	D Lunar Year

68 p.64 Ebenezer Bryce said Bryce Canyon was 'one heck of a place to lose' a what?

A Cow	B Temper
C Fortune	D Leg

69 p.114 What is a caldera?

A Gemstone	B Volcanic crater
C Cold wind	D Light-fitting

70 p.72 Which geological period is named after a Welsh tribe?

A Devonian	B Jurassic
C Silurian	D Carboniferous

Bring me sunshine
All the answers in this round begin with the word 'sun'.

71 p.46 Spectacles with polarising lenses are called what?

72 p.124 Which device displays the time by casting a shadow onto a surface marked with hours?

73 p.46 What 'sun' is a tall yellow annual grown from a small seed?

74 p.46 What do we call dark patches on the Sun's surface caused by cooler areas of gas?

75 p.46 What is the term for a sliding window on the top of a car?

76 p.146 What is the name given to the band of colours caused by sunlight shining through spray?

77 p.124 Which Native American ceremony is performed at the summer solstice?

78 p.46 What was the nickname of Louis XIV of France?

79 p.78 What is the common name for a gem of the mineral feldspar with a reddish-gold sparkle?

80 p.46 Which eight-letter word is another name for strong suntan lotion?

True or false?
Decide if the following statements are correct.

81 p.104 Hong Kong is the world's most populated island.

82 p.64 Humus is decayed plant and animal remains in the soil.

83 p.48 Mercury is the third planet away from the Sun.

84 p.72 All land on Earth was once a single huge continent called Pangaea.

85 p.52 The Moon radiates only 0.01 per cent as much light as the Sun.

86 p.108 Seawater rises and falls four times a day.

87 p.58 Lines of latitude run horizontally around the Earth.

88 p.42 'Quasar' is short for 'quasi-stellar object'.

89 p.58 The Tropic of Capricorn is north of the Equator.

90 p.56 Asteroids that pass close to Earth are called 'Earth-grazers'.

Abbreviations
How well do you know acronyms in use today?

91 p.46 The Sun emits damaging UV rays. What does UV stand for?

92 p.124 GMT is the time at the Prime Meridian line. What does GMT stand for?

93 p.148 Chlorofluorocarbons that damage the ozone layer are better known as what?

94 p.148 What does pH measure?

95 p.148 NO_x is an abbreviation for which dangerous gases?

96 p.112 Divers breathe underwater using SCUBA. What does it stand for?

97 p.152 If something is 400 K, is it very hot, very heavy or very acidic?

98 p.48 When cosmonauts in space make an EVA, what are they doing?

99 p.52 The *Apollo* crew landed on the Moon in an LM. What is it?

100 p.112 What is the term for the echolocation system used to detect objects underwater?

Earthquake!
These questions are all on the subject of earthquakes.

101 p.118 Which fault line along coastal California is named after Saint Andrew?

102 p.118 Which saint, born in Assisi, is said to have healed a boy injured in an earthquake?

103 p.118 What is the surface point directly above the centre of an earthquake?

104 p.118 What is a minor earthquake after the main tremor commonly called?

105 p.118 What happened to the Jamaican coastal town Port Royal after the earthquake of 1692?

106 p.118 The longest recorded earthquake was in Alaska in 1964. Did it last 4 minutes, 40 minutes or 4 hours?

107 p.118 What is the name of the Greek god of earthquakes, who was also god of the sea?

108 p.118 In 1899 in Yakatut Bay, Alaska, an earthquake lifted the coast a record: 5.5 m (17½ft), 11.5 m (37½ft) or 14.5 m (47½ft)?

109 p.118 Which statue, one of the Seven Wonders of the World, was toppled by an earthquake in around 225 BC?

110 p.118 Which naturalist experienced his first earthquake while on a journey around the world on the *Beagle*?

For answers and more facts go to the page given below each question number.
For quick answers to complete quizzes 7 to 13 go to page 154.

It's a picture – movie moments
Ten film-based questions

111 *Crimson* what is a submarine thriller starring Denzel Washington?
p.108

112 Which continent was Meryl Streep 'out of' in 1985?
p.82

113 Who led an elite US military squadron in the movie *The Delta Force*?
p.98

114 In which film does James Bond find Blofeld at the peak of Switzerland's Schilthorn mountain?
p.84

115 In the title of the 1980 rags-to-riches biopic of the country singer Loretta Lynn, what was her father?
p.80

116 Which actor, who played Superman's dad, lives on a private atoll?
p.102

117 Dolph Lungren battles with the evil Skeletor in the sci-fi movie *Master's of the ... what*?
p.40

118 Nick Nolte plays a parent in search of a cure for his son's disease in *Lorenzo's* what?
p.80

119 Barbra Streisand stars alongside Nolte in the film of Pat Conroy's novel *Prince of* what?
p.108

120 In *The English Patient*, the 'patient', Count Almasy, had explored and charted which desert before the Second World War?
p.90

Myths and mysteries
Questions from mythology and past beliefs.

121 Who was the Roman God of the Sea?
p.50

122 Galileo was tried by the Inquisition for challenging which long-held belief?
p.48

123 How was mankind nearly destroyed in the Babylonian poem *Gilgamesh*?
p.122

124 Medieval philosophers believed everything was made from which four elements?
p.40

125 In Egyptian mythology, who was Ra?
p.46

126 According to author Wilkie Collins, which jewel was stolen from an idol's eye and cursed all who possessed it?
p.78

127 Which celestial body is sometimes said to be made of cheese?
p.52

128 Into what was Excalibur thrown while its owner lay mortally wounded?
p.88

129 In which mountains do people claim to have seen the Yeti?
p.84

130 Where is Davy Jones's locker?
p.112

Compare and contrast
Select the correct option from the four possible answers.

131 Which of these rivers is the longest?
p.96

A Amazon	B Nile
C Mississippi	D Volga

132 Which of these regions is the coldest?
p.126

A Arctic	B Siberia
C Alaska	D Antarctic

133 Which of these substances is the hardest?
p.78

A Ruby	B Diamond
C Silicon	D Topaz

134 Which of these cities is nearest the Equator?
p.58

A New York	B Marseilles
C Kiev	D Sofia

135 Which of these oceans is the deepest?
p.112

A Atlantic	B Arctic
C Indian	D Pacific

136 Which of these phases of the Moon is the fullest?
p.52

A New	B Gibbous
C Crescent	D Half

137 Which of these time periods was the longest ago?
p.72

A Jurassic	B Triassic
C Cretaceous	D Carboniferous

138 Which of these countries has the longest coastline?
p.100

A Greenland	B USA
C Chile	D China

139 Which of these planets is farthest from Pluto?
p.48

A Mars	B Saturn
C Jupiter	D Venus

140 Which of the following is farthest below sea level?
p.82

A Death Valley	B Caspian Sea
C Lake Eyre	D Dead Sea

QUIZ 14 WARM UP

Pot luck
A mixed selection of teasers.

141 p.40 What is the name of the astronaut in *Toy Story*?

142 p.80 At what type of power station in Chernobyl did an accident occur in 1986?

143 p.54 Which painter is famous for his 'starry, starry nights'?

144 p.58 What event made William Stukeley write in 1724: 'The earth had lost its blue and was wholly black'?

145 p.146 What, says the legend, lies at the end of a rainbow?

146 p.140 Who is the Norse god of Thunder after whom Thursday is named?

147 p.80 Which fossil fuel is mostly transported around the world by pipeline?

148 p.84 What is the connection between Mount Cervin and the Matterhorn?

149 p.72 Are Alaska and Russia slowly moving apart from or towards each other?

150 p.150 How is the insecticide *Dichlorodiphenyltrichloroethane* better known?

QUIZ 15 IN YOUR STRIDE

Earthy anagrams
Unscramble the anagrams with the help of the clues.

151 p.84 South American mountain range is a sort of SEDAN.

152 p.122 EVINCE an Italian city under threat from severe flooding.

153 p.136 RIVER EROS is a man-made lake used for storing water supplies.

154 p.146 Optical illusion, especially of water in a desert, is a GAMIER.

155 p.90 O BIG desert in southern Mongolia and northern China.

156 p.132 There are NO MOONS in southern Asia's rainy season.

157 p.56 REMOTE TIE from a piece of rock fallen to Earth from outer space.

158 p.90 Ridge of sand formed by the wind and seen in the NUDE.

159 p.84 PREY SEEN on a mountain range between France and Spain.

160 p.120 Mass of snow, ice and rocks sliding down a mountain gives you NAVAL ACHE.

QUIZ 16 ALL COMERS

Also known as...
The answers in this round are all nicknames.

161 p.134 A meteorologically descriptive name for Chicago, Illinois.

162 p.84 A vertebral nickname for the Apennine mountain range.

163 p.56 Another name for a comet: could be thrown in winter by an unwashed child?

164 p.40 An explosive term for the beginning of the Universe.

165 p.78 A name for Ireland, reflecting its green countryside.

166 p.52 Region of Idaho, and National Monument, named for its lunar appearance.

167 p.140 Type of lightning that clings to ships' masts and aircraft wings, as if they were burning.

168 p.82 The hottest, driest basin in North America.

169 p.56 Meteorites as bearers of data about deep space: could be sent to impoverished male astronomers?

170 p.54 Another name for the plough constellation, translated directly from the Latin.

QUIZ 17 CHALLENGE

Name that feature
Give the correct term for the geographical feature.

171 p.66 What do you call a deeply eroded plateau, such as Atherton in Australia?

172 p.72 What name is given to the geological time period from the birth of the Earth to 550 million years ago?

173 p.146 A circle of light around the Sun caused by light refracted through ice crystals is known as a what?

174 p.94 What is the name for a winding river channel caused by alternate bank erosion?

175 p.128 What is the name for the vegetation type between taiga and ice cap?

176 p.90 What do you call a dry watercourse in the desert that contains water only after storms?

177 p.90 What do you call the effect by which the lee side of a mountain range receives far less rain, often creating desert?

178 p.66 When a river cuts down through bedrock, what type of erosion is it doing?

179 p.72 Name the two continents that are believed to have formed after the first great continent, Pangaea, broke up?

180 p.90 What is the term for a desert of bare rock with no sand or gravel?

For answers and more facts go to the page given below each question number.
For quick answers to complete quizzes 14 to 18 go to page 154.

QUIZ 18 WARM UP

Rocks of ages

Which of these rock features is which? Match the pictures to the names below.

- Devil's Marbles
- Uluru
- Grand Canyon
- Cappadocia
- Devil's Tower
- Guilin Hills
- Wave Rock
- Giant's Causeway
- Delicate Arch
- The Mittens

181
p.106

186
p.62

182
p.106

187
p.66

183
p.64

188
p.66

184
p.66

189
p.66

185
p.66

190
p.66

QUIZ 19 WARM UP

QUIZ 20 WARM UP

QUIZ 21 IN YOUR STRIDE

QUIZ 22 ALL COMERS

Water works
All the answers begin with the word 'water'.

191 p.94 Niagara and Angel are both examples of what?

192 p.54 What watery name is given to the constellation Aquarius?

193 p.122 Which scandal forced the resignation of President Nixon?

194 p.134 What is the term for a tornado that whips up a column of water?

195 p.148 In which Kevin Costner film had the ice caps melted, flooding the Earth?

196 p.112 What word describes the level reached by water on the side of a ship?

197 p.136 What is both a turning point in events and a divide between river basins?

198 p.136 What word describes a barrel for collecting and storing rainwater?

199 p.136 What is the process by which water circulates between the oceans, the atmosphere and the land?

200 p.100 Marlon Brando won an Oscar for his part in the 1954 film *On the* what?

Off the literary shelf
Ten questions from the book world.

201 p.50 Where are the invaders from in H.G. Wells' *War of the Worlds*?

202 p.140 Which Shakespearean play features a magical storm whipped up by Prospero?

203 p.132 Which novel contains the words: 'My dear, I don't give a damn!'?

204 p.122 What happened for 40 days and 40 nights in the Book of Genesis?

205 p.128 What is the title of Laura Ingalls Wilder's classic tale of the American West?

206 p.144 *The* what? *Queen* was a haunting tale by Hans Christian Andersen.

207 p.134 In *The Wizard of Oz*, how is Dorothy transported from Kansas to Oz?

208 p.128 In which country is Kipling's *The Jungle Book* set?

209 p.110 *The Old Man and the* what is the title of a book by Ernest Hemingway?

210 p.78 What type of gemstone was described by F. Scott Fitzgerald as being 'as big as the Ritz'?

Sporting chance
The following are all sport-related questions.

211 p.78 Which sport is played on a 'diamond'?

212 p.86 How did Yuichiro Muira descend Mount Everest?

213 p.76 All Olympic Games medals consist almost entirely of which metal?

214 p.52 Alan Shepherd was the first man to play which sport on the Moon?

215 p.50 Which famous tennis sister shares her name with the second planet?

216 p.62 A sheet of what type of rock lies beneath the felt covering on a snooker table?

217 p.142 The broom on which Harry Potter plays Quidditch is named after what type of cloud?

218 p.108 In which sport might you ride a rip curl?

219 p.140 Snooker champion Alex Higgins' quick-playing style earned him what nickname?

220 p.120 In which movie is a ski resort threatened with 'Six million tons of icy terror!'?

Blow by blow
All the answers start with the word 'wind'.

221 p.152 Another name for an anemometer.

222 p.80 A term for an area of land covered by energy-producing wind-turbines.

223 p.92 Two words meaning snow that has been impacted by the wind.

224 p.138 The cooling effect of wind on the Earth's temperature.

225 p.104 A West Indian island group in the path of the north-east trade winds.

226 p.80 A water-sport that uses a sailboard to harness the power of the wind.

227 p.132 A cylinder or cone on a mast showing wind-direction, especially on an airfield.

228 p.152 A means of measuring wind speed, such as the one devised by Sir Francis Beaufort.

229 p.66 A type of valley cut by river erosion through a ridge.

230 p.108 The effect of wind on the sea surface that turns wind energy into waves.

For answers and more facts go to the page given below each question number.
For quick answers to complete quizzes 19 to 25 go to pages 154 and 155.

Pot luck
Challenging teasers on a mixed selection of subjects.

231 p.58
In which hemisphere does the majority of the world's population live?

232 p.116
In which novel does Professor Challenger flee a volcano, while bearing a dinosaur egg?

233 p.112
What is the term for the shallow, gently sloping undersea platform beside a landmass?

234 p.120
What crashed into a reservoir in the Italian Alps in 1963, causing a wave that killed 26 000 people?

235 p.134
What is a chinook?

236 p.66
What was eaten at Hanging Rock in Australia?

237 p.116
When was the last eruption of Vesuvius – 1844, 1874 or 1944?

238 p.94
What is the collective name for the Horseshoe Falls and American Falls?

239 p.88
By what other name is Lac Léman known?

240 p.78
Ninety per cent of the world's opals come from which country?

True or false?
Decide if the following statements are correct.

241 p.148
Plants absorb oxygen and release carbon dioxide.

242 p.40
The farther away a galaxy is, the faster it is moving away from us.

243 p.136
In 1995 an iceberg large enough to supply a fifth of the world's population with water for a year broke away from Antarctica.

244 p.134
Over 95 per cent of forest fires are caused by people.

245 p.148
The average global temperature has risen 1.4°C since 1900.

246 p.124
Seasons occur because the Earth's axis is tilted.

247 p.126
The Sinai Desert and Mount Everest are at the same latitude.

248 p.60
The Earth's crust is thicker beneath the oceans than it is under land.

249 p.130
Concorde flies at up to 18 km (11 miles) above the Earth's surface.

250 p.136
An average bath contains 90 litres (20 gallons) of water.

Mixed meanings
Select the correct option from the four possible answers.

251 p.98
The tidal mouth of a large river is known as a what?

A Actuary	B Emissary
C Estuary	D Ancillary

252 p.128
The upper layer of leaves in a rain forest is called the what?

A Canapé	B Candelabra
C Canopy	D Canasta

253 p.110
The area in the Atlantic Ocean where ships and planes are said to vanish mysteriously is called the Bermuda what?

A Circle	B Diamond
C Square	D Triangle

254 p.58
The obscuring of the Sun by the Moon is a solar what?

A Eclipse	B Ellipse
C Expulse	D Elapse

255 p.150
The impact of a city, mine, factory or quarry on an environment is called its what?

A Footfault	B Footprint
C Footnote	D Footwear

256 p.106
The expanse of swampy land and rivers in southern Florida is called the what?

A Evergreens	B Evermoors
C Everglades	D Evermarshes

257 p.102
What is the word for a group or chain of islands?

A Esperanto	B Amarillo
C Orlando	D Archipelago

258 p.60
What name is given to the pieces of the Earth's crust that move slowly over the mantle?

A Plates	B Dishes
C Cups	D Saucers

259 p.64
The solid layer of rock beneath the soil is called what?

A Rimrock	B Shamrock
C Underrock	D Bedrock

260 p.138
What is the word for the edge of an advancing mass of air?

A Tectonic	B Front
C Cyclone	D Isobar

Take the temperature
Decide whether these places or features are hot or cold.

261 Lava
p.114

262 Comet
p.56

263 Glacier
p.92

264 Mistral
p.134

265 Geyser
p.114

266 Venus
p.50

267 Magma
p.60

268 Floe
p.92

269 Pluto
p.50

270 Atacama
p.90

Star quality
All the answers contain the word 'star'.

271 What is another name for the North Star?
p.54

272 What is the nickname for the oil-rich, second-largest state in the USA?
p.80

273 Which communications satellite was launched in 1962 and immortalised by the pop group The Tornados?
p.44

274 What is the term for a meteor that enters Earth's atmosphere and burns up?
p.56

275 What is the name for a dwarf star that displays spasmodic bursts of radiation?
p.44

276 What is another name for the planet Venus, seen just before sunrise?
p.54

277 What is the term for two stars that revolve round a common centre of gravity?
p.44

278 What is the English name for Sirius, the brightest star in the constellation Canis Major?
p.54

279 What is the name given to any star used to guide the course of a ship?
p.54

280 What type of star emits no visible light, and is also the name of John Carpenter's first sci-fi movie?
p.44

Pot luck
A mixed selection of teasers.

281 Como, Lugano and Maggiore are all lakes in which country?
p.88

282 According to Greek legend, Icarus fell from the sky and drowned in the Aegean Sea. Why did he fall?
p.46

283 What is the common term for pellets of ice that fall from thunder clouds?
p.144

284 What is the town of Évian-les-Bains famous for?
p.136

285 What type of rock is the Giant's Causeway made from?
p.62

286 Who was the first person to cross Niagara Falls on a tightrope?
p.94

287 What is a 'black smoker'?
p.112

288 South America and Africa are on separate continental plates. Are they drifting apart from or towards each other?
p.72

289 What is unusual about Trinidad's Pitch Lake?
p.88

290 What is Norway's main source of power?
p.80

Musical connections
Ten conundrums with a musical theme.

291 'He must know sumpin', but don't say nothin', he jus' keeps rollin.' Who or what is the 'he' in these lyrics?
p.94

292 The film *Pinocchio* featured a song called 'When you wish upon a *what?*'.
p.44

293 What was the name of Bill Haley's backing group?
p.56

294 Which season is it in the famous hit from Gershwin's *Porgy and Bess*?
p.124

295 Mozart's *Symphony Number 41 in C* is named after which planet?
p.50

296 In 1965 Mick Jagger and Keith Richard had a hit called 'Get off of my *what?*'.
p.142

297 'Happy Talk' is a song from which Rodgers and Hammerstein musical?
p.110

298 According to Bob Dylan, 'a hard *what?* is a-gonna fall'.
p.144

299 In the film score of *Breakfast at Tiffany's*, what is 'wider than a mile'?
p.94

300 According to Ira Gershwin, what kind of a day was it in London Town?
p.142

For answers and more facts go to the page given below each question number.
For quick answers to complete quizzes 26 to 33 go to page 155.

Three of a kind
Name the feature that fits all three examples.

301 p.54 Great Bear, Orion, Southern Cross.

302 p.82 Horn, Good Hope, Finisterre.

303 p.110 Barents, Bering, Andaman.

304 p.134 Mistral, föhn, zephyr.

305 p.48 Voyager, Mariner, Venera.

306 p.90 Barchan, star, transverse.

307 p.152 Richter, Mercalli, Moment-magnitude.

308 p.128 Tundra, steppes, chaco.

309 p.50 Io, Europa, Callisto.

310 p.110 Botany, Baffin, Biscay.

All in a name
Provide the names in answer to the following questions.

311 p.84 Which mountain range is named after a god who held up the world?

312 p.106 Which resident of 'Jellystone Park' was named after baseball hero Lawrence Berra?

313 p.56 Which part of the Solar System, named after a Dutch astronomer, is where comets come from?

314 p.116 Which Sicilian volcano takes its name from the Phoenician word for furnace – 'attuna'?

315 p.48 Which cartoon cat had a rock on Mars named after him?

316 p.110 Which canal takes its name from the Arabic *as-suways* – 'the beginning'?

317 p.50 Which planet was named by an 11-year-old schoolgirl after the Roman god of the underworld?

318 p.86 What does the K in K2 stand for?

319 p.58 Which southern line of latitude is named after a sign of the Zodiac?

320 p.152 Which scale is named after a 20th-century seismologist from Ohio?

Quote unquote
Finish or interpret the following quotations.

321 p.142 Complete this line by Winnie the Pooh: 'How sweet to be a *what?* floating in the Blue!'

322 p.124 According to T.S. Eliot, which 'is the cruellest month, breeding Lilacs out of the dead land'?

323 p.48 When first reading Homer, Keats felt 'like some watcher of the skies when a new *what?* swims into his ken'.

324 p.90 In *The English Patient*, what is described as 'a piece of cloth carried by winds, never held down by stones'?

325 p.52 Which US president called the first Moon landing 'the greatest week in the history of the world since the Creation'?

326 p.98 Whose song *Graceland* includes the line: 'The Mississippi Delta was shining like a National guitar'?

327 p.58 Where was Captain Robert Scott when he wrote: 'Great God! this is an awful place!'

328 p.140 According to Shakespeare, what 'quick bright thing unfolds both heaven and earth'?

329 p.144 The poet Shelley wrote of clouds that 'wield the flail of the lashing *what?*'

330 p.52 Where was James Lovell when he said: 'Houston, we have a problem'?

Hard rock
A selection of teasers with a rocky theme.

331 p.62 Which British colony is known as The Rock?

332 p.62 What 'm' is the term for rocks that have changed form?

333 p.62 What 'i' describes rocks that formed from hot, molten material?

334 p.62 How are rocks formed from settled particles collectively known?

335 p.104 In the movie *The Rock*, which prison does the title refer to?

336 p.62 Which rock made from the skeletons of tiny animals is used for writing?

337 p.114 What is molten rock called when it emerges at the Earth's surface?

338 p.62 What colour is Mocha stone?

339 p.62 Which hard stone have humans used for tools, to light fires and to build homes?

340 p.62 What three-word phrase derived from ship-wrecking means 'in a precarious situation'?

QUIZ 34 WARM UP

Land or water?
Are the following features made of land or water?

341 Lagoon
p.100

342 Spit
p.100

343 Loch
p.88

344 Hebrides
p.104

345 Galapagos
p.104

346 Persian Gulf
p.110

347 Rainbow Bridge
p.66

348 Reservoir
p.136

349 Atoll
p.102

350 Sea of Tranquillity
p.52

QUIZ 35 IN YOUR STRIDE

On location
Can you match the locations to the movies listed below?

351 Kennedy Space Center, Florida.
p.52

352 Great Pyramids at Giza, Egypt.
p.96

353 Monument Valley on the Arizona-Utah border.
p.66

354 Amsterdam, Holland.
p.78

355 Devil's Tower, Wyoming, USA.
p.66

356 Mount St Helens, Washington State.
p.116

357 Matmata, southern Tunisia.
p.68

358 Scattergood Nuclear Power Plant, Los Angeles.
p.80

359 Arecibo Observatory, Puerto Rico.
p.44

360 Chattooga River, South Carolina.
p.94

Apollo 13
The China Syndrome
Close Encounters of the Third Kind
Contact
Dante's Peak
Death on the Nile
Deliverance
Diamonds Are Forever
Stagecoach
Star Wars: The Phantom Menace

QUIZ 36 IN YOUR STRIDE

No stone unturned
An assortment of teasers on the theme of rocks.

361 Which soft rock is often used in pottery?
p.62

362 What is the capital of Arkansas, Bill Clinton's home state?
p.62

363 Michelangelo's *David* was sculpted out of which hard stone?
p.62

364 What might 'rocks' mean to a professional thief?
p.78

365 Which gem begins as a grain of sand inside an oyster?
p.78

366 Which porous rock is so light that it will float in the bath?
p.62

367 The Great Pyramid in Egypt is made of which yellow porous stone?
p.62

368 Which very hard rock used for building literally means 'grained'?
p.62

369 What is the word for a small stone made smooth and round by the action of water or sand?
p.66

370 What does a 'rock jock' enjoy doing?
p.86

QUIZ 37 IN YOUR STRIDE

Colour content
All the answers in this round include a reference to colour.

371 What, seen at night, is said to be 'shepherds' delight'?
p.138

372 What colour is the natural volcanic glass obsidian?
p.114

373 When Laurel and Hardy sang 'On the Trail of the Lonesome Pine' which American mountain range did they have in mind?
p.84

374 What are black, white and brown dwarfs?
p.44

375 What is the term for a region of space from which light and matter cannot escape?
p.42

376 Which river is known as 'China's sorrow' because of its fatal flooding?
p.96

377 Which Jamaican mountain range is known for its coffee?
p.84

378 Who, in 1965, was the first US astronaut to walk in space: Edward White, Edward Brown or Edward Green?
p.48

379 Which Italian river was named for its red or ruddy soil?
p.96

380 Transparent ice covering a road surface is referred to as what?
p.144

For answers and more facts go to the page given below each question number.
For quick answers to complete quizzes 34 to 40 go to pages 155 and 156.

Literal names
Match each description to a name in the list below.

381
p.102
Which port and capital is known as the 'Bay of Smokes'?

382
p.64
Which granite buttress in Yosemite National Park is 'the Captain'?

383
p.86
Which extinct volcano in Tanzania is 'the mountain that glitters'?

384
p.94
What is known locally as Mosi-oa-tunya, 'the smoke that thunders'?

385
p.72
Which 'chalky' geological time period stretched from 144 to 65 million years ago?

386
p.60
Which part of the Earth's interior is called a 'ball of stone'?

387
p.66
Which high plateau in southern Colorado is a 'green table'?

388
p.98
Which swamp in Georgia, USA, is named 'trembling earth' after its floating islands?

389
p.72
Which ancient single continent was 'the whole Earth'?

390
p.130
Which layer of the atmosphere is a 'sphere of heat'?

Cretaceous
El Capitán
Kilimanjaro
Lithosphere
Mesa Verde
Okefenokee
Pangaea
Reykjavik
Thermosphere
Victoria Falls

Pot luck
Challenging teasers on a mixed selection of subjects.

391
p.148
Which poisonous gas has the chemical formula CO?

392
p.58
Which three continents does the Equator pass through?

393
p.40
Which unit of measurement represents 9.46 billion km (5.88 billion miles)?

394
p.58
Does the Earth take slightly more or less than 24 hours to rotate once on its axis?

395
p.98
Which country is literally named 'marshland' because 47 per cent of its land is lakes and swamp?

396
p.46
What is the name of the stream of particles given off from the Sun?

397
p.118
What event was Jack London describing when he wrote of San Francisco: 'Nothing remains of it but memories'?

398
p.130
What is measured in millibars?

399
p.110
When the Baltic Sea froze in 1924, it became possible to walk between which two countries?

400
p.40
A parsec is a unit used to measure what?

Just digging it
Select the correct option from the four possible answers.

401
p.70
What does the word 'dinosaur' literally mean?

| A Terrible lizard | B Old Dragon |
| C Big teeth | D Scaly beast |

402
p.70
An evolutionary connection between ape and man is known as a Missing what?

| A Chain | B Fossil |
| C Link | D Sapiens |

403
p.70
An animal or plant turned to stone has been what?

| A Terrified | B Scarified |
| C Petrified | D Ossified |

404
p.70
In *Jurassic Park*, the dinosaur DNA was trapped inside mosquitoes fossilised in drops of what?

| A Blood | B Saliva |
| C Lava | D Amber |

405
p.70
What type of creature does fossil ivory come from?

| A An elephant bird | B A mammoth |
| C A smilodon | D A sabre-toothed tiger |

406
p.70
Which famous fossil find was later shown to be a hoax made from a man's skull and an orang-utan's jaw?

| A Pac-Man | B Piltdown Man |
| C Peat Man | D South Down Man |

407
p.70
Fossilised evidence of, say, a footprint or faeces is a what?

| A Print fossil | B Eco-fossil |
| C Spot fossil | D Trace fossil |

408
p.70
Steinheim 'man', found in 1933, was actually what?

| A A woman | B A dog |
| C A goat | D An orang-utan |

409
p.70
The legend of the one-eyed giant Cyclops may have been inspired by the fossilised skulls of which creature?

| A Ostrich | B Emu |
| C Elephant | D Alligator |

410
p.72
Who thought Earth was created on October 22, 4004 BC?

| A Claudius Ptolemy | B Nicholas Copernicus |
| C Archbishop Ussher | D Pope Gregory XIV |

Go with the flow
Ten questions on rivers around the world.

411 p.96 On which river does the city of Rome stand?

412 p.96 Which river is named after legendary female warriors who were believed to live on its banks?

413 p.96 Which is the longest river in the British Isles?

414 p.96 One river is called Donau in German, Dunaj in Czech, Duna in Hungarian and Dunarea in Romanian. What is its English name?

415 p.96 Which Thai river is best known for the bridge built over it by Allied prisoners of war in World War Two?

416 p.96 Name either of the two rivers that pass through New York City.

417 p.96 Which Agatha Christie novel involves murder on a famous river?

418 p.96 'Teme' was the original Celtic name for which British river?

419 p.96 On which river did Tom Sawyer have his *Adventures*?

420 p.96 Which river in Georgia was celebrated in the song 'Old Folks At Home'?

Literary puzzlers
A selection of teasers on books and writers.

421 p.148 What kind of yellow mist swirled about the London of Sherlock Holmes?

422 p.86 In the Bible, upon which mountain does Noah's Ark land?

423 p.68 The caves of Marabar appear in which novel by E.M. Forster?

424 p.40 Who wrote *A Brief History of Time*?

425 p.56 Mark Twain was born and died supposedly during visits by which celestial phenomenon?

426 p.124 Who wrote the comedy *A Winter's Tale*?

427 p.72 Which novel by Michael Crichton concerns the re-creation of dinosaurs from DNA?

428 p.112 How far 'under the sea' did Captain Nemo go in the Jules Vernes novel?

429 p.58 Which 'shadowy' movie about French poets was scripted by Christopher Hampton?

430 p.104 Who created the buccaneering character Long John Silver?

Counting down
All of the answers involve a number.

431 p.124 Which date occurs just 25 times a century?

432 p.80 On which island did a nuclear accident occur in 1979?

433 p.58 In Jules Verne's novel, how long did Phileas Fogg take to travel around the Earth?

434 p.54 The star cluster Pleiades is also known as The *how many?* Sisters.

435 p.58 In mapping, how many degrees longitude is the Earth divided into?

436 p.108 What percentage of the Earth is covered by the oceans?

437 p.76 How were the Gold-diggers in the 1849 California Gold Rush nicknamed?

438 p.52 How many men have landed on the Moon – 12, 17 or 22?

439 p.84 Ancient Rome was built on how many hills?

440 p.40 Which futuristic frontier space station is the title of a *Star Trek* series?

Pot luck
Challenging teasers on a mixed selection of subjects.

441 p.120 What natural disaster does Pierce Brosnan survive by using an inflatable sphere in *The World Is Not Enough*?

442 p.90 In which US state is the Mojave Desert?

443 p.86 Which cone volcano was a favourite subject of the painter Hokusai?

444 p.56 What is the difference between a meteor and a meteorite?

445 p.84 The Atlas Mountains run through three countries. Name one of the three.

446 p.86 Which mountain overlooks the city of Cape Town?

447 p.132 Who was the ancient Greek god of the north wind?

448 p.64 What type of natural process is ice wedging?

449 p.112 Where on Earth would you find an abyssal plain?

450 p.138 What is an area of low atmospheric pressure commonly called?

For answers and more facts go to the page given below each question number.
For quick answers to complete quizzes 41 to 45 go to page 156.

Spaced out
Can you identify these features in outer space? Match the pictures to the names below.

| Supernova |
| Asteroid |
| Shooting star |
| Nebula |
| Neptune |
| Comet |
| Spiral galaxy |
| Moon's surface |
| Betelgeuse |
| Surface of Mars |

453 p.56

457 p.56

454 p.42

458 p.44

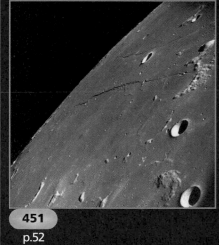

451 p.52

455 p.50

459 p.50

452 p.44

456 p.44

460 p.56

QUIZ 46 WARM UP

Three of a kind
Name a feature or category that fits all three examples.

461 Bondi, Lido, Waikiki.
p.100

462 Small Magellanic Cloud, Milky Way, Andromeda.
p.42

463 Stratus, cumulus, cirrus.
p.142

464 Salyut 1, Skylab, Mir
p.48

465 Jack, hoar, ground.
p.144

466 Leap, light, lunar.
p.124

467 Fungicide, herbicide, insecticide.
p.150

468 Surf, breaker, capillary.
p.108

469 Helios, Hubble, Copernicus.
p.40

470 Black, pot, sink.
p.42

QUIZ 47 IN YOUR STRIDE

Pot luck
A mixed selection of teasers

471 What type of optical illusion is created by light refracting upwards off a hot surface?
p.146

472 Name the seven colours in a rainbow.
p.146

473 Which twice-yearly tides occur when there is the least difference between high and low water?
p.108

474 What is the aboriginal name for the monolith formerly known as Ayer's Rock?
p.66

475 Which weather condition is known as *tai fung* in Chinese?
p.140

476 Which celestial body appears on the Bayeux Tapestry as a bad omen for England's King Harold?
p.56

477 In which country is the Ataturk Dam?
p.80

478 Which ship ran aground in 1989, covering the Gulf of Alaska with oil?
p.150

479 What type of physical feature is an oxbow?
p.94

480 Do rocks expand or contract when they are warmed by the Sun?
p.64

QUIZ 48 IN YOUR STRIDE

Voyages of discovery
Ten questions on explorers and journeys of exploration.

481 How did explorer Amerigo Vespucci leave his mark in atlases around the world?
p.82

482 Where did Viking 1 land in 1976?
p.48

483 Which Norwegian explorer led the first expedition to reach the South Pole in 1911?
p.58

484 Which two oceans are connected by the Strait of Magellan, named after the Portuguese explorer?
p.110

485 At which Australian bay did Captain Cook land in 1770?
p.110

486 Voyager 2 discovered 10 new moons while orbiting which planet?
p.48

487 Which intrepid doctor discovered Victoria Falls?
p.94

488 Pioneer 11 identified two extra rings circling which planet?
p.48

489 Which Caribbean island was named by Christopher Columbus after the Holy Trinity of the Father, Son and Holy Spirit?
p.104

490 How deep is the Mariana Trench: 6 km (3¾ miles), 11 km (6¾ miles) or 16 km (10 miles)?
p.112

QUIZ 49 IN YOUR STRIDE

Starry nights
Match the constellation names to the descriptions.

491 A bird that rose from its own ashes.
p.54

492 A hero who performed 12 labours.
p.54

493 The heavenly Twins, Castor and Pollux.
p.54

494 A bull with brazen feet that was tamed by the hero Jason.
p.54

495 The eagle who carried Jupiter's thunderbolts.
p.54

496 A giant and hunter identified by his club, belt and sword.
p.54

497 A snake whose many heads grew back after they were cut off.
p.54

498 A dolphin who rescued Arion, having been charmed by his song.
p.54

499 A creature with the head, arms and torso of a man and the body and legs of a horse.
p.54

500 A swan into which the god Zeus once transformed himself.
p.54

Aquila
Centaur
Cygnus
Delphinus
Gemini
Hercules
Hydra
Orion
Phoenix
Taurus

For answers and more facts go to the page given below each question number.
For quick answers to complete quizzes 46 to 52 go to page 156.

On song
Ten questions with a musical theme.

501 p.144 When good King Wenceslas looked out, what was deep and crisp and even?

502 p.146 In the song 'Climb Every Mountain', what does Mother Superior tell Maria to follow?

503 p.140 Neil Young sang: 'There's calm in your eye and I'm getting blown away' because you're like a what?

504 p.132 Who did Elton John originally compare to a 'Candle in the Wind'?

505 p.44 What was Lee Marvin 'born under' in the film *Paint Your Wagon*?

506 p.52 According to the pop group The Police, 'giant steps are what you take' when you are walking where?

507 p.78 Which gems are forever, according to Shirley Bassey?

508 p.68 Where does Fingal live in Mendelssohn's Hebridean Overture?

509 p.144 What 'keep fallin' on Butch Cassidy's head?

510 p.104 Which record label took Bob Marley's music to an audience outside Jamaica?

Ten meter dash
All these questions concern measuring instruments.

511 p.152 What does a seismometer measure?

512 p.152 What device measures the imminence of a volcanic eruption ?

513 p.112 A hydrometer measures the density of what?

514 p.130 What type of meter is used to measure atmospheric pressure?

515 p.152 What is measured by an altimeter?

516 p.88 A salinometer would enable somebody to measure what?

517 p.152 What does a pyrometer measure?

518 p.138 What property of air can be established using a hygrometer?

519 p.152 What is measured by an anemometer?

520 p.120 What does a creepometer measure?

Mixed bag
Select the correct option from the four possible answers.

521 p.144 What is the word for droplets of water that condense on cool surfaces at night?

| A Fog | B Frost |
| C Dew | D Ice |

522 p.66 The Grand Canyon was eroded by which river?

| A Colorado | B Rio Grande |
| C Mississippi | D Hudson |

523 p.86 Who, when asked in 1923 why he wanted to climb Mount Everest, replied: 'Because it's there'?

| A Edmund Hillary | B George Mallory |
| C Tenzing Norgay | D Chris Bonnington |

524 p.138 Which line on a meteorological map connects points of the same atmospheric pressure?

| A Millibar | B Milligram |
| C Isobar | D Isotope |

525 p.148 What is the main source of methane in the atmosphere?

| A Cars | B Cows |
| C Volcanoes | D Power stations |

526 p.110 By what other name is La Manche also known?

| A The Spanish peninsula | B The English Channel |
| C The Bay of Biscay | D The Zuiderzee |

527 p.60 The Earth's magnetism is caused by which metal in its core?

| A Iron | B Tin |
| C Uranium | D Copper |

528 p.56 What are the Perseids?

| A A strait | B A Pacific island |
| C A meteor shower | D A high-latitude wind |

529 p.50 Which planet in the Solar System spins in the opposite direction from all the others?

| A Mars | B Pluto |
| C Venus | D Saturn |

530 p.118 How long did the 1906 San Francisco earthquake last?

| A 4.7 seconds | B 47 seconds |
| C 4.7 minutes | D 47 minutes |

All iced up
The answers in this round all begin with the word 'ice'.

531 p.72 A glacial period when ice sheets covered much of the Earth.

532 p.80 A North Atlantic island whose hot springs and volcanoes provide geothermal energy.

533 p.92 A mass of ice attached to land but projecting into the sea.

534 p.92 A large chunk of floating ice hazardous to ships.

535 p.92 A steep part of a glacier resembling a frozen waterfall.

536 p.146 The yellow-white glare in the sky caused by reflection from an ice field.

537 p.92 A vessel with a reinforced bow for cutting a channel through ice.

538 p.92 A thick mass of ice permanently covering the polar regions.

539 p.122 A build-up of ice on rivers that raises the water level, causing floods.

540 p.146 Ice crystals in polar skies that cause haloes and coronas.

Getting about
Ten questions on transport and travel.

541 p.52 The Soviet space probe Luna 9 was the first craft to make a controlled landing where?

542 p.112 What part of the Earth is a bathysphere used to explore?

543 p.58 What was *Arktika* the first ship to reach?

544 p.98 What form of transport has the least difficulty travelling on marshy terrain?

545 p.110 By what means did Ben Abruzzo cross the Pacific in 84 hours 31 minutes?

546 p.100 In which city would you find a *vaporetto*?

547 p.42 Galaxy, Meteor and Vulcan are all types of what vehicle?

548 p.92 What did the SS *Titanic* hit on April 15, 1912?

549 p.92 The *Nautilus* was the first submarine to travel beneath what?

550 p.132 The Ford *Zephyr* was named after the personification of which weather condition?

Mix and match
Unravel the anagrams with the help of the clues.

551 p.122 LINE of a river that floods annually between August and October.

552 p.126 RETRAINED NAME to find a type of climate.

553 p.46 HENRY DOG is the most abundant gas in the Sun.

554 p.106 CHAIN is where you will find the Guilin Hills.

555 p.120 BRIDES find loose rock carried along by landslides or avalanches.

556 p.70 NUNS SAY OUR ART is a large carnivorous dinosaur.

557 p.120 Use OILY MAP to find ancient Olympic Games preserved by mudslides.

558 p.130 RING TONE announces the most common gas in our atmosphere.

559 p.122 NO NEW LASER for southern US city severely flooded in 1965.

560 p.128 I JAIL MARK ON mountain with climate zones ranging from tropical grassland to snow.

Signs and symbols
Name or explain the following signs and symbols.

561 p.76 Cu is the chemical symbol for which element?

562 p.78 What was the name of the biggest diamond ever found?

563 p.52 What are represented on calendars by circles and crescents?

564 p.50 Which planet is sometimes shown by the biological symbol for female?

565 p.76 Au is the chemical symbol for which element?

566 p.152 Which measure of temperature is represented by the letter K?

567 p.76 What is the name of the symbol stamped on gold or silver to certify purity?

568 p.136 What was represented in Egyptian hieroglyphs by a zigzagging line?

569 p.78 Which grass-green gem is the birthstone for May?

570 p.54 Which constellation appears on the Australian flag?

For answers and more facts go to the page given below each question number.
For quick answers to complete quizzes 53 to 57 go to page 156.

High points

Match the mountains and volcanoes in the pictures to the names in the list below.

- Mount St Helens
- Mount Everest
- Table Mountain
- Mount Rushmore
- The Matterhorn
- Mount Fuji
- Popocatépetl
- Heimaey
- Mount Kilimanjaro
- Mount Erebus

571
p.86

572
p.86

573
p.86

574
p.116

575
p.86

576
p.86

577
p.86

578
p.116

579
p.116

580
p.116

QUIZ 58 WARM UP

QUIZ 59 WARM UP

QUIZ 60 IN YOUR STRIDE

QUIZ 61 IN YOUR STRIDE

In other words
Questions on words that have come from other languages.

581 p.48 Which space station's name means 'peace' in Russian?

582 p.140 The word 'blitz' comes from the German for what type of electrical discharge?

583 p.110 Which sea drew its name from the Latin for 'middle of the Earth'?

584 p.100 Now part of the English language, which Norwegian word means a long, narrow inlet?

585 p.108 Which Dutch word, now incorporated into English, begins with 'm' and means a powerful whirlpool?

586 p.98 Which Greek letter is also an area of sediment at the mouth of a river?

587 p.134 Which Italian word beginning with 's' is a hot, dry wind that blows from North Africa to Europe?

588 p.128 What is the Inuit word for 'house'?

589 p.42 What is the Latin word for 'mist', used to describe a cloud of gas in space?

590 p.128 Which Lapp word beginning with 't' describes an area with permanently frozen soil?

Land ahoy!
Answers to the following all begin with the word 'land'.

591 p.120 A term for a slippage of earth or rocks down a cliff or mountain.

592 p.134 A type of light wind that blows towards the sea at night.

593 p.88 Describes a stretch of water surrounded by land, such as the Dead Sea.

594 p.82 A term for a large body of land, such as a continent.

595 p.80 A wind-powered vehicle with sails and wheels used on land.

596 p.82 An object easily seen from a distance and used as a reference point.

597 p.72 A strip of land connecting two continents.

598 p.150 A form of waste disposal.

599 p.102 The peninsula that forms the most westerly point in England.

600 p.150 A series of US satellites that is charting the Earth's resources.

True or false?
Decide if the following statements are correct.

601 p.128 Cacti are only found in American deserts.

602 p.86 Mont Blanc is the highest peak in Europe.

603 p.82 Antarctica is larger than Australia.

604 p.140 A flash of lightning is brighter than 10 million 100-watt light bulbs.

605 p.56 The word 'asteroid' literally means 'like a star'.

606 p.82 Greenland is bigger than India.

607 p.72 England was once joined to the continent of Europe.

608 p.110 The Caribbean is the world's warmest sea.

609 p.60 A magnetic compass always points directly North.

610 p.42 The word 'galaxy' literally means 'milky'.

Hot metal
A series of questions on metals.

611 p.76 If a record sells 600 000 copies, is it classified as 'gold', 'silver' or 'platinum'?

612 p.76 What unit applies both to the purity of gold and the weight of diamonds?

613 p.76 Which metal is also known as 'quicksilver'?

614 p.76 By what common name is the mineral iron pyrite also known?

615 p.76 Which two metals go together to make pewter?

616 p.76 Which metal is used to make the filaments of lightbulbs?

617 p.76 Which is Earth's most common metal: copper, aluminium or iron?

618 p.76 What metal shares its name with a US five cent coin?

619 p.76 What might a 19th-century Californian have used a batea for?

620 p.76 In which Kentucky military base are the US gold reserves held?

For answers and more facts go to the page given below each question number.
For quick answers to complete quizzes 58 to 64 go to pages 156 and 157.

Location, location
Match the places to the films in the list below.

621 Los Angeles, California.
p.118

622 Glacier near Finse, Norway.
p.92

623 North Atlantic Ocean and Mexico.
p.110

624 Kootenai River in Montana, Rogue River in Oregon.
p.94

625 Parkes Radio Telescope, Australia.
p.44

626 Westward Beach, Malibu.
p.100

627 Seville, Spain
p.90

628 Swamps of Louisiana.
p.98

629 Oilfields of Azerbaijan.
p.80

630 Kauai Island, Hawaii.
p.102

The Dish
Earthquake
Jurassic Park
Lawrence of Arabia
Planet of the Apes
  (final scene)
The River Wild
Southern Comfort
Scott of the Antarctic
Titanic
The World Is Not Enough

Colour connections
All the answers include a reference to colour.

631 What are the colours of the five Olympic rings, representing the continents?
p.82

632 The warming of the Earth by gases trapping the Sun's heat is known as the what?
p.148

633 What colour flame is produced by burning natural gas?
p.80

634 What is the term for a wave with a broken crest?
p.108

635 What weather system in Jupiter's atmosphere, measures just over 10000 km (6214 miles) in diameter?
p.50

636 What is sometimes called black gold because of its value?
p.80

637 What is the largest island in the world?
p.104

638 What is the term for a sudden gust of wind bringing rain or sleet?
p.138

639 What is both the name of a large rain forest tree and an award given to members of the US military?
p.128

640 What type of energy radiates from hot objects and can be seen by special cameras?
p.40

Pot luck
Select the correct option from the four possible answers.

641 What is the name for an election victory where one side wins most of the votes?
p.120

A Landlock	B Landmine
C Landrover	D Landslide

642 Where is the Camargue, famous for its wild white horses?
p.98

A Turkey	B France
C Venezuela	D Saudi Arabia

643 What is the nickname for Jigokudani, a region of sulphurous hot springs in Japan?
p.114

A Death Valley	B Tornado Alley
C Valley of Hell	D Happy Valley

644 What is the term for a short-lived but intense flood?
p.122

A Flash flood	B Flush flood
C Fresh flood	D Froth flood

645 What colour is the mineral albite?
p.74

A Orange	B Purple
C White	D Green

646 What mineral is the basic ingredient of glass?
p.74

A Fluorite	B Silica
C Borax	D Gypsum

647 What does the Latin word *fossilis* mean?
p.70

A Very old	B Turned to stone
C Dug up	D Judge

648 An exposed, weathered body of rock is an exfoliation what?
p.64

A Spire	B Dome
C Peak	D Tower

649 What is a fumarole?
p.114

A A cave of ice	B A type of comet
C An Italian island	D A vent in a volcano

650 What triggered an avalanche in the resort of Val D'Isère on February 12, 1999?
p.120

A Oil spillage	B Car backfiring
C Meteorite	D Snowboard race

In their own words
Finish or interpret the following quotations.

651 p.124 What did Keats call the 'Season of mists and mellow fruitfulness'?

652 p.40 Which scientist said: 'Politics is for the present but an equation is something for eternity'?

653 p.142 In Wordsworth's poem, what 'floats on high o'er vales and hills'?

654 p.52 Who said in 1969: 'The surface is fine and powdery – I can pick it up loosely with my toe'?

655 p.58 What was first seen from the far side of the Moon in 1968 as 'small and beautiful in that eternal silence where it floats'?

656 p.114 Complete the missing word in Havelock Ellis's description of civilisation: 'a thin crust over a ____ of revolution'?

657 p.74 What is the missing word in this Gilbert and Sullivan lyric: 'In matters vegetable, animal, and ____, I am the very model of a modern Major-General.'

658 p.124 What, in Rodgers and Hammerstein's *Carousel*, 'is bustin' out all over'?

659 p.52 According to the poet Ted Hughes, what 'rolls along the hills, gently bouncing, A vast balloon, till it takes off, and sinks upward'?

660 p.150 Which pop star said: 'If I were a Brazilian without land or money … I would be burning the rain forest too'?

Mountains high
Ten questions with a mountain theme.

661 p.86 The heads of four US Presidents are carved into the side of which mountain?

662 p.84 Which city in Monaco has a name that means 'Charles' Mountain'?

663 p.86 Which peak did Clint Eastwood climb in a film with the mountain's name in its title?

664 p.86 Which mountain was the legendary home of the Greek gods?

665 p.84 In which mountain range was *The Sound of Music* set?

666 p.84 Mussorgsky's *Night on Bald Mountain* was animated in which 1940 Disney classic?

667 p.84 Which US state has a name that literally means 'mountainous'?

668 p.86 Who starred as a mountaineer in the film *Vertical Limit*?

669 p.106 In which US state are the Teton mountains?

670 p.86 According to the Bible, on which mountain did Moses receive the Ten Commandments?

Three of a kind
Name a category or feature that fits all three examples.

671 p.80 Crude, refined, sunflower.

672 p.150 Intensive, arable, dairy.

673 p.114 Extinct, dormant, active.

674 p.70 *Diplodocus*, *Triceratops*, *Oviraptor*.

675 p.120 Creep, earthflow, lahar.

676 p.106 Yosemite, Serengeti, Kruger.

677 p.98 Bog, swamp, marsh.

678 p.58 Elliptical, geo-stationary, low-Earth.

679 p.42 Spiral, barred spiral, elliptical.

680 p.68 Batmanhole, Sistema, Pierre-Saint-Martin.

Pot luck
A mixed selection of teasers.

681 p.124 Does the Earth circle the Sun more slowly during a leap year?

682 p.124 Where can you experience sunlight 24 hours a day?

683 p.132 Which wind beginning with 'm' ushers in a season of heavy rain?

684 p.60 What is the uppermost structural layer of the Earth commonly called?

685 p.128 Which is bigger: Europe or the Amazon rain forest?

686 p.126 In March 1911, what covered Tamarac, California, to a depth of 11.46 m (37 ft)?

687 p.122 What is the word for an embankment built to prevent flooding, especially flooding by the sea?

688 p.120 In nature, what type of event is a sluff?

689 p.120 A lahar is a mudflow created when ash mixes with water or melted snow during what kind of event?

690 p.44 What is the term for stars that emit regular pulses of radio waves?

For answers and more facts go to the page given below each question number.
For quick answers to complete quizzes 65 to 71 go to page 157.

What's it called?
Identify or explain the names referred to in each question.

691 p.84 Which country is sometimes referred to as The Roof of the World?

692 p.40 The Big Bang was the Universe's explosive beginning. What is the nickname for the theory that it will eventually shrink and collapse?

693 p.138 By what 'forecasting' name is the Scarlet Pimpernel flower also known?

694 p.54 Which constellation is sometimes called the winged horse?

695 p.48 Which four planets in our Solar System are collectively known as the gas giants?

696 p.148 Which city earned the nickname Big Smoke while blighted by heavy industry in the 1800s.

697 p.46 What nickname for Japan is reflected in its national flag?

698 p.92 On which continent is there a US base known as Little America?

699 p.134 In West Africa, what type of weather condition has the nickname The Doctor?

700 p.150 What is the common term for clearing forest land for farming?

And now, the weather
Ten questions on weather and weather-forecasting

701 p.130 What is inflated then released to collect weather data?

702 p.126 In what type of climate do subpolar and subtropical air masses alternate?

703 p.130 What measures one kilogram per square centimetre at sea level around the world?

704 p.138 What forms when a cold front overtakes a warm front in a low-pressure area?

705 p.134 What 't' are vertical winds caused by the Sun warming some areas of land more than others?

706 p.138 Which five-letter weather term was coined in and inspired by the First World War?

707 p.138 If it 'rains before seven', what will it be 'by eleven'?

708 p.138 What is the technical term for a map showing general weather conditions over a particular area?

709 p.138 What do weather forecasters estimate by combining temperature and wind speed?

710 p.138 What do triangles along a line on a weather map indicate?

Scientific studies
Select the correct option from the four possible answers.

711 p.136 Hydrology is the study or science of what?

A Caves	B Energy
C Hurricanes	D Water

712 p.152 What is the study of earthquakes called?

A Seismology	B Shudderology
C Quakology	D Volcanology

713 p.78 Gemology is the science of what?

A Gemstones	B Forests
C Metals	D Lakes

714 p.138 What do meteorologists study?

A Meteors	B Metals
C Asteroids	D Weather

715 p.40 Cosmology is the study or science of what?

A Wind	B Volcanoes
C Universe	D Climate

716 p.64 What is studied in the science of agrology?

A Soil	B Seas
C Ice	D Clouds

717 p.130 Aerology is the science of which of the following?

A Bubbles	B Radiation
C Atmosphere	D Climate

718 p.70 What are palaeontologists experts on?

A Planets	B Fossils
C Caves	D Deserts

719 p.88 Limnology is the study of what?

A Stars	B Snow
C Weather	D Lakes

720 p.62 Which of the following are studied by petrologists?

A Rivers	B Fossil fuels
C Rocks	D Soils

Down to Earth
All the answers begin with the word 'earth'.

721 p.118 A violent shaking of the ground, caused by movement of plates.

722 p.46 The glow caused by sunlight reflected by the Earth onto the Moon.

723 p.66 A column of rock topped by a boulder that protects it from erosion.

724 p.150 The UN-based system for monitoring environmental activities.

725 p.150 A dam built by compacting successive layers of earth.

726 p.120 A sheet of water-soaked soil and rock that slips slowly down a slope.

727 p.60 The study of the structure and age of the Earth.

728 p.118 Another name for the god Neptune.

729 p.52 The appearance of the Earth above the horizon of the Moon.

730 p.150 An environmental sculpture, such as Robert Smithson's *Spiral Jetty* in Utah's Great Salt Lake.

Pot luck
A mixed selection of teasers.

731 p.60 What is the centre of the Earth usually called?

732 p.122 Which US river burst its banks in 1993, causing the country's worst-ever flooding?

733 p.92 The International Ice Patrol was established in 1914, following which disaster?

734 p.58 Who is said to have discovered gravity after seeing an apple fall from a tree?

735 p.150 What were destroyed in 1943 in order to flood the Ruhr valley?

736 p.90 In which desert would you find Africa's Bushmen?

737 p.46 Which star is monitored by the satellite Solar Max?

738 p.68 What is the difference between a cave and a pothole?

739 p.148 Which is more destructive, acid fog or acid rain?

740 p.44 What is the word for an exploding star that glows brilliantly bright for a short while?

Island games
The following questions are on the theme of islands.

741 p.102 By what name are the Malvinas in the South Atlantic better known?

742 p.104 In *Gulliver's Travels*, what is special about the island of Laputa?

743 p.102 What is the name of the legendary island where King Arthur is supposed to be buried?

744 p.104 Who wrote 'No man is an island, entire of it self.'

745 p.104 Who was exiled on the island of Elba off the Italian coast?

746 p.104 The novel *Papillon* was set on which prison island off French Guiana?

747 p.104 Who was stranded on a desert island in Daniel Defoe's most famous novel?

748 p.104 What is the largest island in the Mediterranean Sea?

749 p.102 In which island group are Fuerteventura and Lanzarote?

750 p.104 Which island in the Pacific is famous for its giant statues of heads?

Every one counts
Either the question or the answer includes a number.

751 p.56 Asteroid 2309 is named after which officer in *Star Trek*?

752 p.122 How long did the biblical flood last?

753 p.144 How many points does a snowflake usually have?

754 p.46 How long does light from the Sun take to reach the Earth: 4, 8 or 12 minutes?

755 p.74 The Mohs scale measures the hardness of minerals from 1–10. What number is 'talc'?

756 p.124 How many time zones is the Earth divided into?

757 p.132 What is the name for the area of the southern oceans characterised by strong westerly winds and wet weather?

758 p.120 Who lost 18000 men, 2000 horses and several elephants in an avalanche?

759 p.124 Who composed the *Four Seasons*?

760 p.68 How tall, to the nearest 10m (30ft), is the world's tallest stalagmite?

For answers and more facts go to the page given below each question number.
For quick answers to complete quizzes 72 to 76 go to page 157.

QUIZ 76 CHALLENGE

Coast to coast

Each map shows a section of coastline from one of the places listed below. Can you match up the maps with the names? (The maps are not all to the same scale.)

Italy
Alaska
Mexico
Somalia
England
Chile
Australia
Canada
India
Hawaii

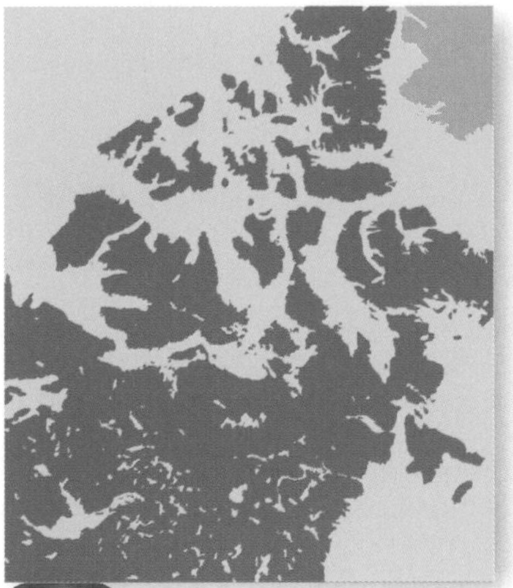

761
p.100

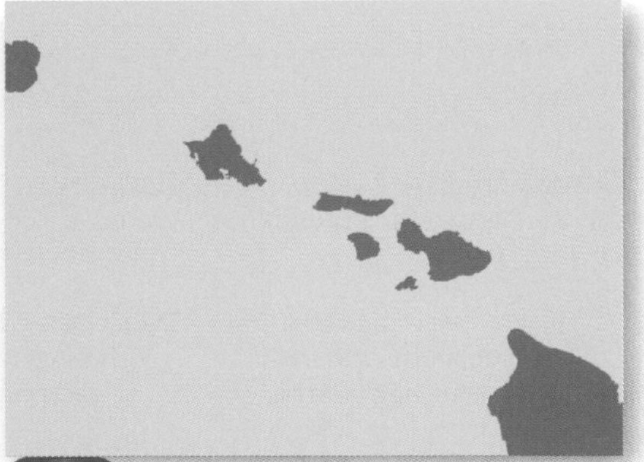

762
p.100

763
p.100

764
p.100

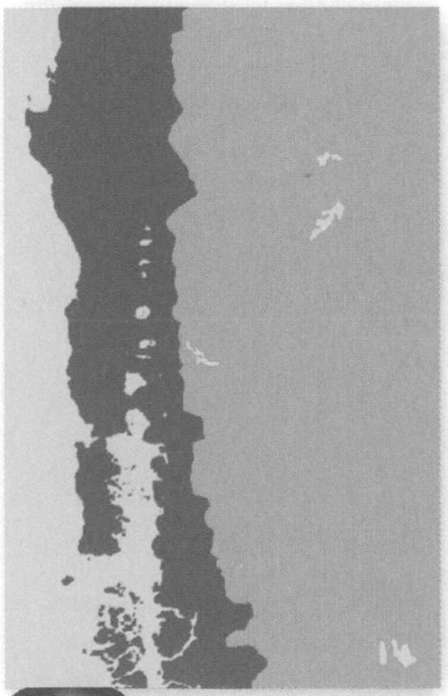

765
p.100

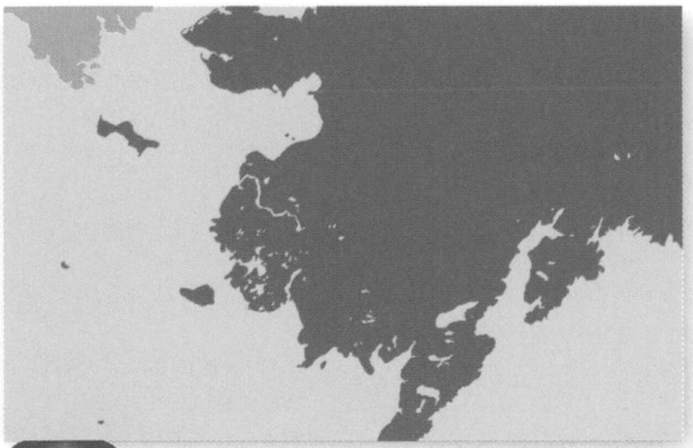

766
p.100

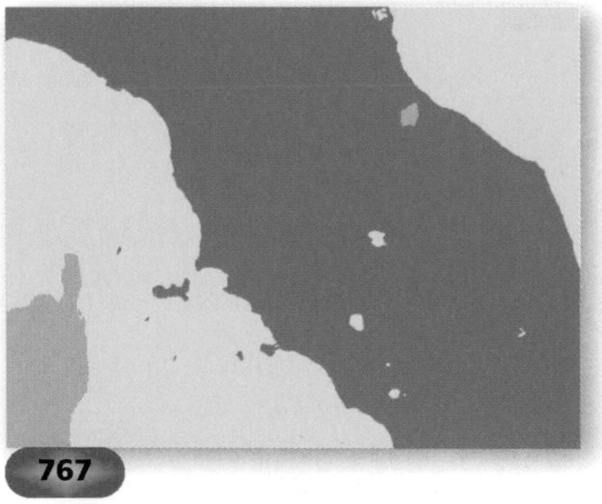

767
p.100

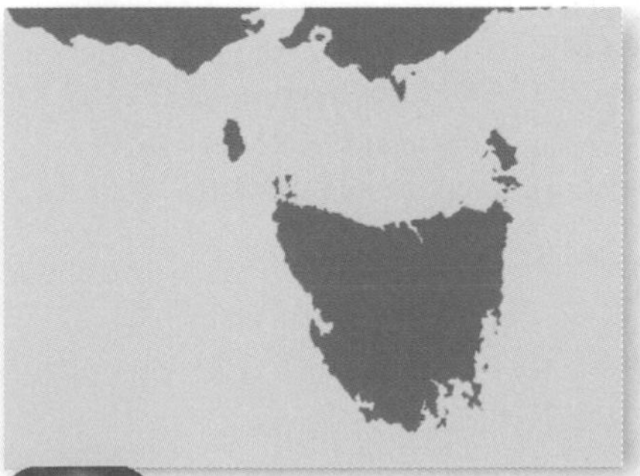

768
p.100

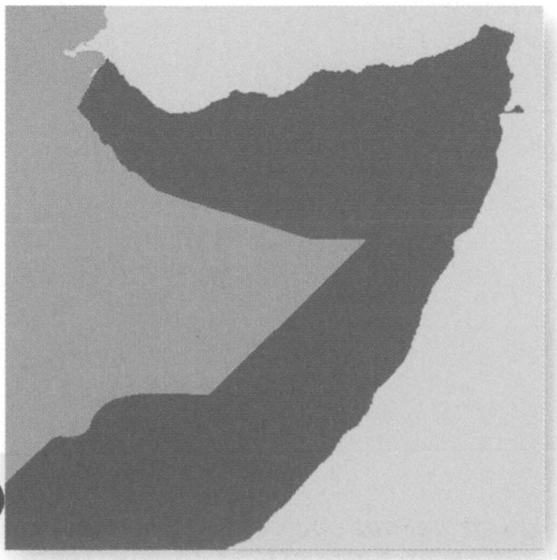

769
p.100

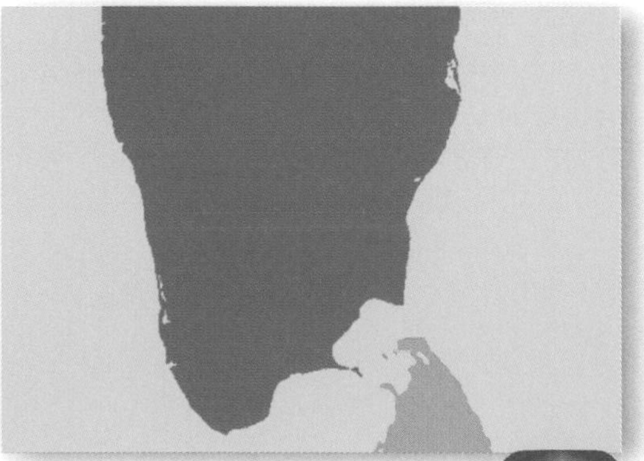

770
p.100

Down by the lake
Ten questions on the theme of lakes.

771 p.88 Lakes Erie, Huron, Michigan, Ontario and Superior are known collectively as what?

772 p.88 Lake Geneva lies on the border between which two countries?

773 p.88 Which film, Henry Fonda's last, was filmed on Squam Lake, New Hampshire?

774 p.88 Crater Lake in Oregon, USA, lies inside the remains of what?

775 p.88 Which lake, on the border of Peru and Bolivia in the Andes, is the world's highest navigable lake?

776 p.88 Which East African lake was named by the explorer John Speke after a British queen?

777 p.88 What were William Wordsworth, Robert Southey and Samuel Taylor Coleridge known collectively as?

778 p.88 Where in Utah did the Mormons found a city in 1847?

779 p.88 Where does Nessie, the legendary monster, live?

780 p.122 Lake Nasser in Egypt was created in 1970 through the building of what?

Musical medley
A selection of questions with a musical connection.

781 p.132 According to Bob Dylan, where is the answer blowing?

782 p.94 Where did a 'little tom-tit' sing 'Willow, tit-willow, tit-willow'?

783 p.124 Name the 1963 film in which Cliff Richard drove a London bus?

784 p.128 Complete the missing word in this line from 'God Bless America': 'From the mountains to the _____ to the oceans white with foam'.

785 p.140 In the musical *Grease* what electrical phenomenon do the cast sing about?

786 p.144 Prince released an album in 1984 called *Purple* what?

787 p.140 Tom Jones sang the theme to which Bond movie?

788 p.56 Goodness Gracious! How did 1950s rocker Jerry Lee Lewis describe meteors?

789 p.140 According to *My Fair Lady*, which phenomena 'hardly happen' in 'Hertford, Hereford and Hampshire'?

790 p.52 What part of the Moon does Pink Floyd's classic 1970s album refer to?

Three of a kind
Name a feature or category that fits all three examples.

791 p.82 Africa, Asia, Europe.

792 p.80 Solar, hydroelectric, nuclear.

793 p.44 Alpha Centauri, Sirius A, Betelgeuse.

794 p.106 Cloud, rain, petrified.

795 p.130 Ozone, boundary, active.

796 p.44 Jodrell Bank, Parkes, Arecibo.

797 p.62 Gabbro, coquina, shale.

798 p.126 Tropical, temperate, polar.

799 p.120 Creep, debris flow, rockfall.

800 p.60 Asteroid, van Allen, green.

Four corners
The answers all relate to points of the compass.

801 p.102 New Zealand is divided by the Cook Strait into which two islands?

802 p.58 Which Pole lies at the Antarctic?

803 p.122 The Thames flood barrier protects London from flooding by which sea?

804 p.86 Which Hitchcock film featured a spectacular sequence on Mount Rushmore?

805 p.86 Which is the most difficult face of the Eiger to climb?

806 p.96 Does the Amazon River flow mainly north-east or south-west?

807 p.108 The Humboldt Current flows along the west coast of: North America or South America?

808 p.132 In which direction do the north-east trade winds travel?

809 p.60 What does a magnetic compass needle point towards?

810 p.50 On Earth, the Sun rises in the east. In which direction does it rise on Venus?

For answers and more facts go to the page given below each question number.
For quick answers to complete quizzes 77 to 83 go to pages 157 and 158.

Lights, camera, action
Ten questions with a movie connection.

811 p.116 What is Dante's Peak in the 1997 Pierce Brosnan film of the same name?

812 p.78 In the 1964 Peter Sellers movie, *The Pink Panther*, what is the Pink Panther?

813 p.122 Which 1998 film about a reptilian monster features a huge wave?

814 p.94 What 'runs through' Robert Redford's movie about fly-fishing?

815 p.76 Which 1987 Stanley Kubrick movie is about life in the US Marines at the time of the Vietnam War?

816 p.56 What threatens the Earth in the 1998 film *Armageddon*?

817 p.48 The 1956 science fiction film, *Forbidden Planet*, in which Dr Morbius and his daughter are stranded on a remote planet, is based on which play by Shakespeare?

818 p.152 Which François Truffaut film is named after the temperature at which paper catches fire?

819 p.46 Which 1950 movie, starring Gloria Swanson, opens with a corpse lying face down in a swimming pool?

820 p.140 Which 1983 film starring Roy Scheider features a 'silent' police helicopter?

Volcano alert
Answer these questions on a volcano theme

821 p.116 Where was the 1997 film *Volcano* set?

822 p.114 Vulcan was the Roman god of what?

823 p.114 What 'c' is a crater formed when the centre of a dormant or extinct volcanic cone falls inwards?

824 p.116 Which novel by Jules Verne is about a voyage that begins on Mount Etna?

825 p.50 Which character invented by Saint-Exupery has two volcanoes on his planet?

826 p.114 In vulcanology, what is meant by 'repose' time?

827 p.116 Arthur's Seat, an old volcano eroded to a hill, overlooks which city?

828 p.116 Which two towns were destroyed when Vesuvius erupted in AD79?

829 p.116 In which film does Tom Hanks agree to jump into a volcano?

830 p.50 Three times higher than Mount Everest, Olympus Mons is a volcano on which planet?

Pot luck
Select the correct option from the four possible answers.

831 p.58 What imaginary line lies at 0° latitude?

| A Greenwich Meridian | B Tropic of Cancer |
| C International Dateline | D Equator |

832 p.74 What type of mineral would you find in a pencil?

| A Calcite | B Potassium |
| C Graphite | D Sulphur |

833 p.150 Which country is planting a 'Great Wall' of trees to counter erosion from deforestation?

| A Bangladesh | B China |
| C Brazil | D Italy |

834 p.48 How many planets are there in the Solar System?

| A Seven | B Nine |
| C Ten | D Eight |

835 p.82 Which continent is sometimes called the cradle of mankind?

| A Australia | B North America |
| C Africa | D Asia |

836 p.76 Which strong, light metal is used in aircraft manufacture?

| A Titanium | B Strontium |
| C Lead | D Silver |

837 p.90 In which country is the Negev Desert?

| A Russia | B Israel |
| C Tunisia | D Australia |

838 p.60 What is the layer of molten rock beneath Earth's surface called?

| A The plateau | B The mantle |
| C The shield | D The shelf |

839 p.48 Planets beyond the Solar System are referred to as what?

| A Extrasolar | B Intergalactic |
| C Supernumerary | D Dark |

840 p.150 Where was the 1992 Earth Summit held?

| A Kyoto | B Rio de Janeiro |
| C Brussels | D Cape Town |

QUIZ 84 WARM UP

Red, white or blue?
Fill in the blanks with either 'red', 'white' or 'blue'.

841 p.144 A Christmas with snow on the ground is a _____ Christmas.

842 p.50 Mars is also known as the _____ Planet.

843 p.94 A stretch of fast-flowing water used for rafting is called _____ water.

844 p.78 The commonest form of a precious gemstone used in jewellery is the _____ sapphire.

845 p.106 A _____ wood is an evergreen tree that can grow to over 100 m (330 ft) tall.

846 p.74 Found in bloodstones, the mineral jasper is _____.

847 p.130 The scattering of the Sun's rays against the dark background of space makes the sky appear _____.

848 p.124 A night in summer in the Arctic Circle when the Sun never sets is a _____ night.

849 p.96 Strauss's waltz about the second longest river in Europe is called the _____ *Danube*.

850 p.46 In a few billion years' time the Sun will turn into a _____ Giant.

QUIZ 85 IN YOUR STRIDE

Under the Earth
Ten questions with a cave connection.

851 p.68 What are the caves at Lascaux in France famous for?

852 p.68 Which magic words did Aladdin use to enter the thieves' cave?

853 p.68 Do stalagmites grow upwards or downwards?

854 p.68 Which superhero keeps his car in a cave?

855 p.68 The Mammoth Cave system in Kentucky is the longest in the world. Is it: 225 km (140 miles), 555 km (345 miles), or 750 km (466 miles) long?

856 p.68 What substance covers the surfaces of the Eisriesenwelt cave system in the Austrian Alps?

857 p.68 What is unusual about the cave-dwelling fish *Amblyopsis rosae*?

858 p.68 Which ancient texts were found in caves in the Judaean wilderness in the mid-20th century?

859 p.68 What type of cheese was first made in a Somerset village famous for its gorge and caves?

860 p.68 What is the study of caves called: speleology, campanology or subteratology?

QUIZ 86 ALL COMERS

Building bridges
Identify these famous bridges or their locations.

861 p.98 Alistair MacLean's story *Golden Gate* is set on the bridge over which bay?

862 p.94 The Queenston-Lewiston Bridge is downriver from which waterfall?

863 p.96 Which river does Tower Bridge span?

864 p.96 Pont Neuf spans which river?

865 p.122 In which Italian city is the Ponte Vecchio?

866 p.110 The Akashi-Kaikyo Bridge crosses which sea?

867 p.96 Which river does the George Washington Bridge span?

868 p.88 The bridge now crossing Lake Havasu in Arizona was formerly in which city?

869 p.96 Which US river does the Mark Twain Memorial Bridge span?

870 p.96 In German legend, a bridge of gold crossed which river at Bingen?

QUIZ 87 CHALLENGE

True or false?
Decide if the following statements are correct.

871 p.90 The Sahara Desert covers almost the same size area as the USA.

872 p.126 Local variations in conditions at ground level are called microclimates.

873 p.122 Tsunamis occur most frequently in the Pacific Ocean.

874 p.44 New stars are born in 'nurseries'.

875 p.94 All rivers end up eventually in an ocean or sea.

876 p.148 The main world source of pollution inside homes is the burning of fossil fuels.

877 p.120 Eighty per cent of Alpine avalanches occur between January and March.

878 p.118 Most earthquakes originate on a landmass.

879 p.40 Our galaxy has been expanding ever since the Big Bang.

880 p.126 The coldest inhabited place in the world is in Siberia.

For answers and more facts go to the page given below each question number.
For quick answers to complete quizzes 84 to 88 go to page 158.

QUIZ
88
CHALLENGE

Turned out nice again
Can you identify these weather conditions and atmospheric phenomena? Match the pictures to the names listed below.

- Cirrus cloud
- Cumulus cloud
- Cumulonimbus cloud
- Lenticular cloud
- Crespuscular rays
- Tornado
- Glory
- Aurora
- Sun halo
- Sun pillar

881
p.146

882
p.146

883
p.142

884
p.134

885
p.146

886
p.142

887
p.146

888
p.142

889
p.146

890
p.142

89 WARM UP

Three of a kind
Name a feature or category that fits all three examples.

891 Halley's, Hale-Bopp, Shoemaker-Levy 9.
p.56

892 Forked, sheet, ball.
p.140

893 Polar, pack, fast.
p.92

894 Urals, Andes, Pyrenees.
p.84

895 Suez, Panama, Grand.
p.110

896 Hipparchus, Herschel, Galileo.
p.48

897 Trilobites, ammonites, belemnites.
p.70

898 Baltic, Tasman, South China.
p.110

899 Nubian, Gobi, Kalahari.
p.90

900 Belize Barrier, Kingman, Great Barrier.
p.102

90 IN YOUR STRIDE

Pot luck
A mixed selection of teasers.

901 Which tailless amphibians rained down on England in 1939?
p.144

902 How thick is the atmosphere, to the nearest 100 km (60 miles)?
p.130

903 What is the name of the layer of partially decomposed rock beneath the topsoil and above the bedrock?
p.64

904 When do icebergs usually 'calve' – in summer or winter?
p.92

905 What 'h' means water in the air?
p.138

906 What fire-resistant mineral fibre has been found to cause lung cancer?
p.74

907 What thins away to nothing over the Antarctic every spring?
p.148

908 What fictional mineral does Superman fear?
p.74

909 What is the term for a low-lying area of land by a river that often bursts its banks?
p.94

910 Where would you find a magma chamber?
p.114

91 IN YOUR STRIDE

Delta connection
Select the correct river delta from the list given below.

911 Cairo lies on the _____ delta.
p.98

912 Arles lies on the _____ delta.
p.98

913 My Tho lies on the _____ delta.
p.98

914 New Orleans lies on the _____ delta.
p.98

915 Shanghai lies on the _____ delta.
p.98

916 Ostia lies on the _____ delta.
p.98

917 St Petersburg lies on the _____ delta.
p.98

918 La Esperanza lies on the _____ delta.
p.98

919 Astrakhan lies on the _____ delta
p.98

920 Osaka lies on the _____ delta.
p.98

Chang Jiang
Mekong
Mississippi
Neva
Nile
Orinoco
Rhône
Tiber
Volga
Yodo

92 IN YOUR STRIDE

All the world's a stage
A series of questions with a theatrical connection.

921 What type of weather is imitated in theatres by shaking a large sheet of metal?
p.140

922 Robert Bolt's play about Thomas More is called *A Man for All* what?
p.124

923 *The Admirable Crichton* is a butler who becomes king when stranded on a what?
p.102

924 Complete the missing word in this line from *Richard III*: 'Now is the winter of our discontent made glorious _____ by this sun of York.'
p.124

925 Ben Jonson's *The Alchemist* strives to turn base metal into what substance?
p.76

926 In what type of theatre are 'sky shows' projected onto the inside of a dome?
p.54

927 In *The Birds*, Aristophanes talks of an imaginary city-state in the air. What is this city called?
p.142

928 Why was San Francisco's Opera House renovated in 1996?
p.118

929 Which Roman city, destroyed by a volcano, was revealed by the discovery of an ancient theatre?
p.116

930 Coleridge once famously described watching Edmund Kean's acting as being 'like reading Shakespeare by flashes of' – what?
p.140

For answers and more facts go to the page given below each question number.
For quick answers to complete quizzes 89 to 95 go to page 158.

QUIZ 93 ALL COMERS

QUIZ 94 CHALLENGE

QUIZ 95 MULTIPLE CHOICE

Full palette
Either the question or answer has a reference to colour.

931 p.88
What has turned the water pink in East Africa's Lake Natron?

932 p.144
What is the term for a heavy frost without snow or rime?

933 p.110
The Atlantic Ocean was first crossed in an aeroplane by Alcock and who?

934 p.78
What colour is the copper carbonate mineral malachite?

935 p.96
Which African river was named after the Dutch royal family?

936 p.106
The presence of which metal gives rock a green or blue colour?

937 p.110
Which sea separates the East Coast of China from the Korean peninsula?

938 p.80
What readily available source of power has the nickname white coal?

939 p.112
What is the largest living thing in the ocean?

940 p.74
Hellfire and lightning were once believed to be made from which yellow element?

Great ideas
Questions on inventors and their inventions.

941 p.152
What did Hans Geiger's invention enable him to detect and measure?

942 p.58
What did John Harrison's marine chronometer enable sailors to measure?

943 p.152
What nationality was astronomer Anders Celsius, inventor of the Celsius temperate scale?

944 p.112
What did Cornelius Drebbel invent in 1620 that later furthered the exploration of the oceans?

945 p.112
Which famous undersea explorer invented the aqualung?

946 p.44
Spectacle-maker Hans Lippershey called his invention in 1608 a kijker or 'looker'. What was it?

947 p.46
Which invention by Frenchman Eugene Schueller protects us from harmful rays?

948 p.74
The Chinese invented which explosive combination of charcoal, saltpetre and sulphur?

949 p.80
Whose splitting of the atom in 1919 paved the way for nuclear energy?

950 p.140
What protective device for buildings was invented by Benjamin Franklin?

Wind power
Select the correct option from the four possible answers.

951 p.138
Complete the proverb: 'No weather is ill if the wind is ...'

A Still	B Chill
C Brill	D Up hill

952 p.132
The usual wind direction in any one place is the what?

A Promising wind	B Presiding wind
C Providing wind	D Prevailing wind

953 p.138
Complete: 'When the wind is in the West, the weather is ...?

A At its best	B Having a rest
C Depressed	D A right pest

954 p.134
What is the term for a rising current of warm air?

A Dermal	B Diurnal
C Thermal	D Ancillary

955 p.132
A 'westerly' wind blows from which direction?

A West	B East
C North	D South

956 p.132
A subtropical high-pressure belt encircling the Earth is known as a what latitude?

A Dog	B Horse
C Cow	D Chicken

957 p.134
Where does a hot summer wind called a 'brickfielder' blow?

A Argentina	B Sahara Desert
C Antarctica	D Australia

958 p.132
What is the term for the strong wind that blows about 10km (6 miles) above the Earth?

A Jet stream	B Cold stream
C Gulf stream	D Wind stream

959 p.134
The name *Simoom* refers to a hot, dry Arabian wind. What does the term literally mean?

A Rattlesnake	B Dead dog
C Poison wind	D Simon's storm

960 p.126
On Antarctica's George V Coast, the world's windiest place, winds blow at speeds of up to _____?

A 520km/h (323mph)	B 120km/h (74mph)
C 720km/h (447mph)	D 320km/h (199mph)

QUIZ 96 WARM UP

A pinch of salt
A set of questions on the theme of salt.

961 p.88 — The Caspian Sea is the world's largest lake. Is it filled with fresh or saltwater?

962 p.74 — What 'b' is another word for saltwater?

963 p.74 — In the Bible, whose wife was turned into a pillar of salt?

964 p.112 — What percentage of saltwater is salt: 3 per cent, 6 per cent or 9 per cent?

965 p.136 — What name is given to the process of removing salt from seawater?

966 p.74 — What word for payment derives from the salt paid to Roman soldiers?

967 p.112 — What is the chemical term for table salt?

968 p.74 — By what name is halite better known: rock salt or sea salt?

969 p.112 — Which contain more salt: warm seas or cold seas?

970 p.74 — Does salt raise or lower the boiling point of water?

QUIZ 97 IN YOUR STRIDE

The naming of things
Questions on names and their origins.

971 p.110 — Which ocean was named by the explorer Magellan because he found it 'peaceful' when he sailed into it?

972 p.50 — Which planet is named after the Roman god of science and travellers?

973 p.140 — Which Atlantic weather conditions will be named Alberto, Beryl, Chris and Debbie in 2006?

974 p.84 — Which 'white mountain' has the highest peak in the Alps?

975 p.146 — Which phenomenon caused by solar particles hitting the atmosphere has a name that literally means 'Dawn of the North Wind'?

976 p.104 — Which icy country was given a pleasant name to encourage settlers to go there?

977 p.148 — Which two words are mixed together to form the word 'smog'?

978 p.142 — What type of cloud has a name that means 'pile' in Latin?

979 p.132 — Which area of the ocean, once dreaded for becalming ships, came also to mean a period of depression?

980 p.108 — Which phenomenon linked to ocean currents has a Spanish name that translates as 'The Boy Child'?

QUIZ 98 ALL COMERS

Playing out
All of these questions have a musical connection.

981 p.142 — The pop group Lindisfarne sang about what on the River Tyne?

982 p.144 — What was 'cruel' in the Christmas carol 'Good King Wenceslas'?

983 p.78 — In the film *Gentlemen Prefer Blondes*, what 'are a girl's best friend'?

984 p.146 — In *The Wizard of Oz*, Dorothy went to a place that was 'somewhere over the' what?

985 p.132 — What was the 1995 Oscar-winning song from *Pocahontas* called?

986 p.52 — By what name is Beethoven's Piano Sonata No.14, Op 27 No 2, better known?

987 p.84 — 'River Deep *what*? High' was a hit for Ike and Tina Turner.

988 p.68 — In 1965, Bob Dylan sang about his what 'Homesick Blues'?

989 p.72 — What will there 'be bluebirds over' in 'There'll Always be an England?'

990 p.124 — A riot occurred at the premiere of which work by Stravinsky?

QUIZ 99 CHALLENGE

How many? How far?
Match the statements to the figures in the list below.

991 p.126 — The lowest temperature, in degrees Celsius (°C), ever recorded on Earth.

992 p.84 — The number of metres Tibet has risen in the last 2 million years.

993 p.58 — The number of days that the Earth takes to orbit the Sun.

994 p.120 — The number of people killed between 1975 and 1989 by avalanches in the European Alps.

995 p.152 — On the Richter Scale, the highest measurement ever recorded for an earthquake.

996 p.94 — The height in metres of the highest waterfall in the world.

997 p.42 — The approximate number of stars in the Milky Way.

998 p.46 — The percentage of the Sun's mass made of hydrogen and helium.

999 p.42 — The speed in kilometres per hour at which the Milky Way is moving through the Universe.

1000 p.72 — The age of the Earth in years.

200 billion
4.6 billion
2 172 150
3000
1622
807
365.24219
98
9.5
−89.2

For answers and more facts go to the page given below each question number.
For quick answers to complete quizzes 96 to 99 go to page 158.

PLANET EARTH

AND THE

UNIVERSE

Essential facts, figures and
other information on our planet and
its place in the Solar System

The Universe

Core facts ❶

◆ The **Universe** encompasses everything – planets, stars, galaxies and space.
◆ The most popular current theory on the origin of the Universe states that space and time began 15 billion years ago with the **Big Bang** and that the Universe has been expanding ever since.

◆ The whole Universe is made of the **same elements** that are found on Earth.
◆ **Cosmology** is the study of the Universe – its origins, future and how it functions – by the analysis of **electromagnetic radiation** from space objects.
◆ No one knows how, or if, the Universe will end.

The Big Bang ❷

According to the Big Bang theory, the Universe began from a 'singularity', a tiny point, smaller than a single atom, that exploded in a **massive fireball** which then expanded at unimaginable speed. A fraction of a second after this Big Bang, **subatomic particles** formed.

As the Universe cooled over the next billion years, the subatomic particles reacted with background radiation from the Big Bang, **clumping together** into hydrogen and helium atoms.

After 2 billion years, the **first stars** and galaxies began to condense from these gas clouds.

★ 124

Elements, atoms and the Ancients

Medieval thinkers, influenced by the Ancient Greek philosopher, Aristotle, believed that everything was made of **four elements**: air, water, earth and fire. In fact, two earlier Greek thinkers, Leucippus and Democritus, had come closer to the truth. They proposed that everything was made up of indestructible particles – or **atoms** – too small to be seen or subdivided. They reasoned that these atoms and space were all that existed and that all change came from the atoms rearranging themselves.

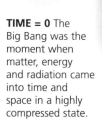

TIME = 0 The Big Bang was the moment when matter, energy and radiation came into time and space in a highly compressed state.

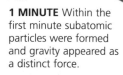

1 MINUTE Within the first minute subatomic particles were formed and gravity appeared as a distinct force.

Electromagnetic spectrum ❸

Objects in space give off **radiation**; electric and magnetic disturbances that travel through space as waves of varying length. This radiation is known as the electromagnetic spectrum and ranges from gamma waves, the shortest wavelength, through X-rays, ultraviolet, visible light, infrared, microwaves and radio waves. Cosmologists analyse the radiation emitted by objects in **deep space** to work out what they are, what they are made of and their distance from Earth.

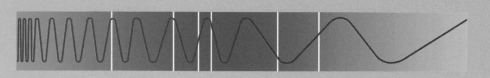

An expanding Universe ❹

In 1929 US astronomer Edwin Hubble proposed the theory that the Universe is expanding. He based this on observations of light emitted by distant galaxies. The wavelengths of light from an object lengthen as it moves away from the observer, causing the light to shift towards the red end of the spectrum – **redshift**. Hubble measured the redshift of more than 24 galaxies, and discovered that the farther away a galaxy is, the faster it is receding. In the past galaxies must have been closer together. But although the Universe is expanding, galaxies themselves can either stay the same size, expand or contract.

Measuring space ❻

The vastness of space requires large units of measurement. Astronomical units are used for distances within our Solar System, and parsecs and light years for distances between stars and galaxies.
Astronomical Unit (AU) = 150 million km (93 million miles), equal to the distance from the Earth to the Sun.
Light year = 9.46 billion km (5.88 billion miles), the distance travelled by light in one year. Light travels at just under 300 000 km/sec (186 000 miles/sec).
Parsec = 31 billion km (19 billion miles), or 3.26 light years.

HUBBLE VISION ❺
The Hubble Space Telescope, launched in 1990, has allowed us to see farther into space than ever before. It can perceive objects 4 billion times fainter than the human eye can discern.

Where are we? ❼

The Universe is thought to contain billions of galaxies. They are collected together in groups and clusters, which in turn are grouped in **superclusters**. Between them are vast empty regions. It is not known whether galaxies formed first and clumped together into clusters, or whether clusters formed first and then split up into galaxies.

1 Our Milky Way **Galaxy** belongs to a small subgroup, called the **Local Group**, on the edge of the **Virgo Supercluster**.

2 The **Milky Way Galaxy** is just one of over 30 galaxies in the Local Group.

3 Our **Solar System** is about two-thirds of the way out from the centre of the Milky Way Galaxy.

4 Earth is the third planet out from the Sun in our Solar System.

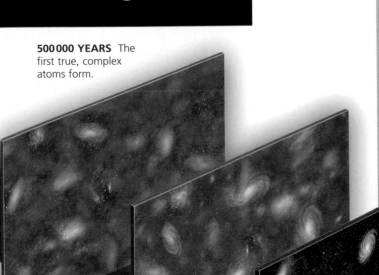

500 000 YEARS The first true, complex atoms form.

1 BILLION YEARS Clouds of matter start forming under the effect of gravity.

2 BILLION YEARS The first stars and galaxies begin forming from clouds of gas.

The future of the Universe ❽

There are several theories about the eventual fate of the Universe.
◆ It may expand for ever, eventually simply fading away.
◆ It may stop expanding and stay as it is.
◆ It may reach an optimum size and then slowly collapse under the force of gravity – the Big Crunch theory.
◆ It may go through alternate expansion and contraction phases for ever.

QUESTION NUMBER

The numbers or star following the answers refer to information boxes on the right.

ANSWERS

59   *Galaxy Quest* – Star Trek spoof made in 1999

67   **A: Cosmic year** ❻

★ **88**   **True** ★

375   **Black hole** ❶ ❸

454   **Spiral galaxy** ❶ ❷

462   **Galaxy** ❼

470   **Hole** ❸

547   **Aeroplane** – they are all military aircraft

589   **Nebula** ❺

610   **True** – from Greek *galaxias kuklos*, milky circle/way

679   **Galaxy** ❷

997   **200 billion** ❻

999   **2 172 150 km/h (1 346 733 mph)**

Galaxies

Core facts ❶

◆ **Galaxies** are collections of billions of stars held together by the gravitational pull they exert on each other. They evolved from clumps of matter distributed irregularly through space.
◆ Around **100 billion** galaxies are currently known to exist in the visible Universe.
◆ Galaxies are **classified** as either **normal** or **active**, based on the size and rate of activity of the black holes at their centres.

◆ There are **four types** of normal galaxies: spiral, barred spiral, elliptical and irregular.
◆ The **Milky Way** is a spiral galaxy, and our Sun and Solar System lie on one of the spiral arms.
◆ Some galaxies belong to small groups, others occur in large clusters of hundreds or thousands of galaxies. **Superclusters**, which can reach diameters of up to 100 million light years, contain tens of thousands of galaxies.

Normal galaxies ❷

These are classified by their overall shape. **Spiral galaxies** have arms wound round a central nucleus. **Barred spirals** have arms spiralling round a bar-shaped core. Both contain a high level of star-forming activity. **Elliptical galaxies** are oval and have no arms. They may have formed from collisions between spiral galaxies. They show little star-forming activity, and contain mainly old stars. **Irregular galaxies** have no specific shape or structure.

Barred spiral galaxy NGC 1530 has arms radiating out from a central bar-like structure.

Classical spiral galaxy NGC 6946 is a galaxy in the constellation of Cepheus.

Irregular galaxy NGC 55, a galaxy in our Local Group.

Quasars ★ 88

Quasars are the **super-bright cores** of galaxies that are otherwise invisible. They are the brightest objects in the Universe, giving off thousands of times more light than a normal galaxy. The luminosity is thought to be energy radiated by matter as it is pulled into **supermassive black holes** at the centres of the quasars. These black holes are many billion times the mass of the Sun. Six thousand quasars have been detected, all from the early Universe. Some are thought to have been formed from collisions between neighbouring galaxies.

Galactic black holes ❸

Most galaxies are thought to have central black holes – **collapsed stars** with such strong gravitational fields that even light cannot escape from them, which is why they cannot be seen directly.
   Active galaxies have large black holes that pull in a constant stream of gas and dust clouds and star fragments. There is probably a relatively small black hole – just 100 000 times the mass of the Sun – at the centre of our galaxy.

Tie-breaker ❹

Q: Which famous astronomer was also an amateur boxing champion?
A: Edwin Hubble. He boxed at the University of Chicago in 1906-10; some saw him as a possible contender for world heavyweight champion.

Milky Way Galaxy ❺

◆ Our Galaxy consists of four main components: the central bulge, the disc of spiral arms, the halo and globular clusters. There may also be an outer halo, the galactic corona. The disc rotates, but the central bulge does not.

◆ The central bulge and globular clusters contain mainly old stars, known as Population II stars, which formed from original cosmic matter.

◆ The spiral arms, where new star formation takes place, contain mainly middle-aged and young stars, known as Population I stars. These formed from recycled star matter, and are metal rich.

HALO
Spherical region surrounding the disc.

CENTRAL BULGE/NUCLEUS
Contains mainly old stars, dust and gas.

GLOBULAR CLUSTERS
Clusters of old stars. About 150 globular clusters are distributed through the galactic halo. Each one contains many millions of stars.

SPIRAL ARMS
Streams of bright stars circle the galaxy centre. Most, including our Sun, are young and middle-aged stars. Clouds of dust and gas, called nebulae, where new stars form, are interspersed through the spiral arms.

Milky Way Galaxy: vital statistics ❻

Age 12–14 billion years

Total diameter 80000–100000 light years

Depth at centre 6000 light years

Disc depth at Sun's location 1000 light years

Number of stars At least 200 billion

Location of Sun On the Orion arm, two-thirds of the way out from the centre (30000 light years)

Speed of orbit of Sun and Solar System around the centre 250km/sec (155 miles/sec)

Time Sun takes to complete one orbit 225 million years, known as a cosmic year

THE MILKY WAY Seen from Earth, our Galaxy appears as a band of stars across the night sky.

The Local Group ❼

The Milky Way's Local Group consists of three large and 30 small galaxies. **Andromeda**, 165000 light years in diameter, is the **largest galaxy** in the Group. It contains at least 400 billion stars, and is the only galaxy that can be seen with the naked eye in the Northern Hemisphere. Our **nearest galaxies** are the Large and Small Magellanic Clouds 160000 and 190000 light years away, and the recently discovered Sagittarius Dwarf Galaxy around 80000 light years away.

ANDROMEDA The largest galaxy in our Local Group and closest spiral neighbour.

WEIRD AND WONDERFUL ❽
Dark matter does not give off radiation so cannot be detected. Galaxies may be embedded in dark matter whose gravitational pull helps to hold them together, despite the gravitational fields of neighbouring galaxies.

Stars

Core facts ❶

◆ **Stars are** balls of gas, composed mainly of hydrogen and helium and held together by gravity, in which nuclear fusion creates energy.
◆ A **star's evolution** depends on its mass. The larger the star, the more gravity compresses the material within it, the hotter it becomes, and the more rapidly nuclear reactions take place.

◆ **Star lifespans** vary from a few million years to thousands of millions of years.
◆ When **stars die**, such as in **supernova** events, elements are recycled to form new stars and solar systems.
◆ Stars may appear as single points of light, but many are made up of two or more stars.

Evolution of stars ❷

A star's life cycle is divided into three main phases, driven by the conflict between the **outward** pressure of heat and energy supporting a star and the **inward** tug of gravity. Some stars fail to achieve nuclear fusion but still radiate some light. They are known as brown dwarves.

1 BIRTH
Thousands, even millions, of stars form together in 'nurseries' inside dust and gas clouds (nebulae).

Dust and gas are drawn together by gravity.

A protostar continues contracting and heating up. As the nebula disperses, it continues in its own orbit.

2 MAIN SEQUENCE
The star is hot enough to sustain nuclear reactions in its core.

3 DEATH
This begins when hydrogen in the core runs out. It takes one of three courses depending on the star's size.

A **medium star** expands into a red giant, cools and shrinks into a white dwarf, then cools further to form a black dwarf.

A **large star** explodes as a supernova, then collapses under its own gravity into a **neutron star**. Some neutron stars rotate.

A **giant star** explodes as a supernova, then collapses completely, forming a **black hole**.

Hotter (younger)   Cooler (older)

Brighter (larger)

Duller (smaller)

Hertzsprung-Russell diagram

This graph is used to work out a star's type and age. The star's surface temperature (calculated from the colour of the light it emits) is plotted against its brightness, and the star is classified according to where it falls on the graph. **Main sequence** stars, such as our Sun, fall in the band across the middle.

★ 740

BANG! The spectacular and short-lived explosion of star 1987A into a supernova.

Supernovae – stellar recyclers

A supernova is the explosive death of a giant star. The last one seen was in 1987 in the Large Magellanic Cloud. Elements heavier than iron, such as gold, are created in the intense heat of a supernova. They are scattered as interstellar dust, from which new stars and planets form.

Variations ❹

Pulsar A spinning **neutron** star that sends out a regular pulse of radio waves.

Binary star Two stars, held together by gravity, that circle around each other. Optical binaries are stars in the same line of sight that appear linked but are actually separate.

Variables Stars that display fluctuating brightness.

Telescopes ❸

Different types of telescope are used to study the stars:
◆ **Optical telescopes** use either lenses to refract or mirrors to reflect distant objects. The world's largest reflecting telescope is in Zelenchukskaya, Russia. Its biggest mirror weighs 70 tonnes and is 6 m (20 ft) wide.
◆ **Radio telescopes** use giant dishes – or an array of linked dishes – to pick up radio waves. At 305 m (1000 ft) wide, Arecibo in Puerto Rico is the world's largest dish telescope.
◆ **Space telescopes** placed in orbit observe stars without interference from the Earth's atmosphere.

LISTENING TO THE STARS
The massive dish of the Parkes radio telescope in NSW, Australia.

Well-known stars ❺

Star	Claim to fame
Pistol Star	Brightest known star (10 million times more powerful than the Sun)
Sirius A	Brightest-looking star in night sky, 24 times brighter than the Sun. Also known as the Dog Star.
Proxima Centauri	Nearest (4.23 light years). Next nearest are Alpha Centauri A and B
Betelgeuse	One of the largest known (diameter 1000 million km/620 million miles)
Shurnarkabtishashutu	Longest name. Arabic for 'under the southern horn of the bull'
Hyades (300 stars)	Nearest star cluster, c.150 light years away in Taurus constellation
PSR B1937+214	Fastest-spinning pulsar (642 revolutions per second)

WEIRD AND WONDERFUL ❻

The first researchers to spot pulsars gave them the nickname **LGMs** (Little Green Men), because these stars emit signals so regular and precise, it is as if they were produced by an intelligent alien life-form.

Star songs ❼

Musicals invoking the stars include: *Singin' in the Rain* (1952), with Gene Kelly and Debbie Reynolds singing 'You Are My Lucky Star'; *Pinocchio* (1940), with 'When You Wish Upon A Star'; and *Paint Your Wagon* (1969) with Lee Marvin singing 'Wand'rin' Star'.

The Sun

Core facts   **1**

◆ The Sun is a **medium-sized star** formed 4.6 billion years ago. It has used up around half of the hydrogen fuel in its core. In about 5 billion years it will expand to become a red giant, engulfing Mercury and Venus and possibly Earth.
◆ **Nuclear reactions** in the Sun's core send waves of energy flowing to the surface, where it is given off as light, heat and other radiation.

◆ The Sun's **diameter** is 1 392 000 km (865 000 miles). More than 1 million Earths would fit inside the Sun.
◆ The Sun is the **source of heat and light** for the whole Solar System. Its light takes 8.3 minutes to reach Earth.
◆ **Violent activity** such as sunspots, flares and prominences are seen on the Sun's surface.

What is the Sun made of?   **2**

The Sun is composed of 71 per cent **hydrogen**, 27 per cent **helium**, and 2 per cent **heavier elements** such as carbon, oxygen, nitrogen, neon and iron.

The extreme heat and pressure in the Sun causes hydrogen atoms to fuse into helium, releasing energy. The particles that transport heat energy (called photons) follow a 'random walk' from the centre to the surface of the Sun by a combination of convection and radiation. The transfer of heat energy from the core to the surface can take up to 10 million years.

WEIRD AND WONDERFUL **3**
The Sun has 11-year cycles of **sunspot activity** due to changes in its magnetic field. The increased surface activity means that more charged solar particles reach Earth's atmosphere, causing radio interference and aurorae.

Inside the Sun   **4**

Around 5 million tonnes of the 700 million tonnes of hydrogen the Sun burns every second is converted into pure energy. This process also produces 695 million tonnes of helium per second. The core is surrounded by two inner layers, the radiative layer, and the convective layer. Above these is the visible surface, called the photosphere, and two further layers of atmosphere, the chromosphere and the corona.

THE CORE The Sun's nuclear reactor where hydrogen fusion occurs.

INTERIOR The convective layer and, beneath it, the radiative layer lie below the visible surface.

ATMOSPHERE The photosphere, the Sun's visible surface, is surrounded by two outer layers, the chromosphere and the corona.

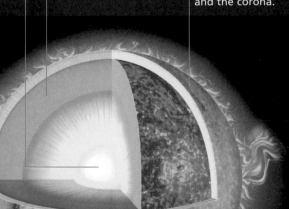

The Sun's radiation ⑤

◆ Most of the Sun's radiation is in the ultraviolet, visible light and infrared wavelengths.
◆ Sunlight is necessary for plant photosynthesis.
◆ Heat, in the form of infrared radiation, creates the mean global temperature necessary for life, and provides the energy for oceanic and atmospheric circulations.
◆ Most of the Sun's harmful ultraviolet radiation is blocked by the ozone layer, but it still causes sunburn, cancers and cataracts.

INDIAN SUMMER The sun rises over Vembanad Lake, Southern India.

Surface features ⑥

Sunspots Dark patches on the surface can be up to 100000 km (62000 miles) in diameter. Strong magnetic fields in these areas inhibit the flow of energy to the surface, so they are cooler than the surrounding area. Sunspots last for between one hour and one month.
Spicules Flame-like columns of gas that shoot up to 10000 km (6200 miles) from the surface.
Faculae Temporary bright spots that appear on the Sun's surface.
Prominences Flame-like loops supported by the Sun's magnetic field that rise tens of thousands of kilometres. When seen against the Sun's surface they appear dark and are called filaments.
Solar flares Explosive releases of energy eject clouds of atomic particles into space, triggering microwave and radio-wave radiation. These can cause electrical interference on Earth, affecting television screens and computers and creating surges in power lines.

ACTIVE SUN Solar flares and sunspots are characteristic of the active phase of solar cycles.

PROMINENCES The arcs of charged gas known as prominences are relatively cool compared to the surface of the Sun.

Solar wind

★ 396

The solar wind is a continuous flow of charged atomic particles that stream from the Sun's surface at speeds of up to 1000 km/sec (600 mps). The fastest streams come from holes in the corona, the Sun's outermost layer.

Sun gods ⑦

The Babylonian sun god was Shamash; for the Persians it was Mithras. The Egyptian god Ra was born in the sky each morning and died an old man each night. For ancient Romans, Phoebus Apollo rode a fiery chariot across the sky. The Aztec sun gods Tezcatlipoca and Huitzilopochti demanded human sacrifice, while the Japanese sun-goddess Ameratasu is reflected in Japan's national symbol.

SUN GOD RA The Egyptian deity and father of the gods.

Solar System 1

Core facts ①

◆ The Solar System **consists** of the Sun, the nine planets in orbit around the Sun, 77 satellites – or moons – orbiting the planets, asteroids, comets, and all other debris circulating around the Sun.
◆ The Solar System **originated** as a cloud of dust and gas.

◆ The Sun accounts for 99 per cent of the Solar System's mass; most of the rest is in Jupiter.
◆ Everything in the Solar System is held in orbit by the **Sun's gravity**.
◆ The planets are divided into **inner**, terrestrial ones and **outer**, gaseous ones.
◆ **Extrasolar planets** may exist beyond Pluto.

MERCURY Small rocky planet, no moons.

VENUS Similar in size to Earth, no moons.

EARTH Only planet known to support life. One moon.

MARS Cold red surface, two moons

ASTEROID BELT A band of rocky debris that failed to form a planet.

Formation ②

Around 4.5 billion years ago a cloud of dust and gas pulled together by gravity began rotating, and material at its centre heated up to form the proto-Sun. Other circulating matter accreted together to form the planets, which orbit the Sun in the same direction and the same plane.

JUPITER The largest planet. It has 16 moons.

Inner and outer planets ③

◆ Mercury, Venus, Earth and Mars are classified as **inner terrestrial**, or rocky, planets. They are made mainly of rock and metal, have solid surfaces, and rotate slowly.
◆ Jupiter, Saturn, Uranus and Neptune are classified as **outer gas giants**. They have deep atmospheres of mainly hydrogen and helium, and they rotate quickly.
◆ **Temperatures** on Mercury and Venus are too high to support life. On Mars it is too cold.

SATURN A gas giant with rings of suspended ice particles and 30 moons.

URANUS Uniquely in the Solar System, it rotates on its side. It has 18 moons.

WEIRD AND WONDERFUL ④

Planet V was a fifth terrestrial planet that lay just beyond Mars. About 3.9 billion years ago, it was pulled out of orbit by the gravity of its neighbours, disrupting the asteroid belt before being pulled into the Sun and destroyed.

NEPTUNE Faint rings, a methane-rich atmosphere and eight moons.

PLUTO Smallest planet. Its orbit sometimes brings it closer to the Sun than Neptune. One moon.

Exploration of the planets ❺

Unmanned space probes have visited every planet except Pluto. Some probes are orbiters; others are landers.

Launched	Mission	Event
1962	**Mariner 2 (USA)**	First probe to return data on Venus
1965	**Mariner 4 (USA)**	Sent back the first pictures of Mars
1969	**Mariner 6 and 7 (USA)**	Examined Mars' atmosphere
1970	**Venera 7 (USSR)**	First landing on Venus; the first controlled landing on another planet
1972	**Venera 8 (USSR)**	Landed on Venus and transmitted messages
1973	**Mariner 10 (USA)**	First fly-by of Mercury
1973	**Pioneer 10 (USA)**	First fly-by of Jupiter
1973	**Pioneer 11 (USA)**	First fly-by of Saturn
1975	**Viking 1 and 2 (USA)**	First landing on Mars (in 1976); sent back images and soil analysis
1977	**Voyager 1 and 2 (USA)**	Explored Jupiter's moons; went on to Uranus and Neptune
1992	**Ulysses**	Fly-by of Jupiter's north and south poles
1995	**Galileo (USA)**	First probe into Jupiter's atmosphere
1997	**Pathfinder (USA)**	Sojourner rover made chemical analyses of rocks on Mars

THE EYE OF JUPITER
This photograph of the massive, swirling permanent storm in Jupiter's atmosphere, known as the Great Red Spot, was taken by the Galileo probe in 1995.

PATHFINDER ON MARS
The Sojourner rover meets a rock nicknamed Yogi Bear.

⭐ 122

Galileo

The theory of an **Earth-centred Solar System** was put forward by the Greek astronomer Ptolemy. It was first challenged by Nicolaus Copernicus (1473–1543), who believed that the planets orbited the Sun. This controversial view was proven by Galileo (1564–1642) a century later. Galileo was imprisoned for his heresy and his books banned.

Extrasolar planets ❻

◆ In 1991 the first planets outside our Solar System were discovered. Three planets, similar in mass to Earth, were found **orbiting a pulsar**.
◆ In 1999, the first planet **orbiting a star** similar to our Sun was found in the constellation Pegasus, 50 light years from Earth. The Hubble telescope and ground-based telescopes studied its atmosphere and found that it is a large gas planet, similar to Jupiter and Saturn. It takes just 3.5 days to orbit its parent star.
◆ Altogether, around 80 planets have so far been discovered outside the Solar System.

Star films ❼

The most successful space movies ever made were the *Star Wars* series. The original trilogy – *Star Wars* (1977), *The Empire Strikes Back* (1980) and *Return of the Jedi* (1983) – together grossed US$1.9 billion worldwide. *Episode I: The Phantom Menace* (1999) and *Episode II: Attack of the Clones* (2002) brought the total up to more than $3.37 billion.

Solar System 2

Mercury

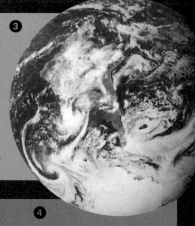

Mercury is the planet **closest to the Sun** at 45.9–69.7 million km (28.5–43.3 million miles). Daytime temperatures reach 425°C (800°F). It has a solid iron core and little atmosphere. It is heavily cratered by meteorites: its largest crater, the Caloris Basin, is 1300 km (800 miles) across. It **orbits the Sun faster** than any other planet, at 172 248 km/h (107 030 mph).

MERCURY Named for the Roman god of travel, Mercury orbits the Sun in just 88 days.

Venus

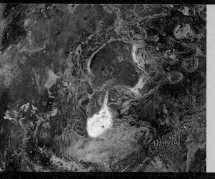

Similar to the Earth in size and mass, Venus has a thick, poisonous atmosphere that conceals its surface and traps heat. Temperatures rise to over 480°C (900°F), making it the Solar System's **hottest** planet. It is also the **brightest** as its atmosphere reflects back 70 per cent of sunlight. It rotates slowly backwards, so the Sun rises in the west. Its day lasts 243 Earth days – the Solar System's **longest day**.

MAXWELL MOUNTAINS The highest on Venus at 11 km (7 miles).

Earth

The only planet in the Solar System – or as yet in the Universe – known to **support life** is protected from the Sun's radiation by an atmosphere of mainly nitrogen and oxygen. Its rocks are composed predominantly of iron, oxygen, magnesium, silicon and nickel. It is the only planet that has **surface water** (70 per cent), so it looks blue from space. Earth's solar orbit takes 365.242 days, and it is tilted on its axis at an angle of 23.5°.

THE BLUE PLANET Earth has constantly changing cloud cover.

Mars

Named after the Roman god of war, Mars is also known as the Red Planet because of the high level of red iron oxide in its soil. Its day is almost the same length as Earth's. Surface features include Olympus Mons, the **largest volcano** in the Solar System, 600 km (375 miles) wide and 25 km (15 miles) high, and Mariner Valley, 13 times longer than the Grand Canyon. Mars has two tiny moons, Phobos (Fear) and Deimos (Panic).

ROCKY SURFACE Mars photographed by Viking 1.

Tie-breaker

Q: Which planet's moons are named after Shakespearian characters?
A: Uranus. The five largest are Titania, which has ice cliffs and fault lines on the surface, Oberon, Ariel, Umbriel and Miranda. Puck has a mottled surface. Cordelia and Ophelia escort Uranus' outermost ring.

Jupiter ❻

Named after the Roman king of the gods, Jupiter is a world of extremes. It is the **largest** planet in the Solar System, 1300 times bigger than Earth – all the other planets would fit into it twice over. It is the **fastest-spinning** planet at 45500 km/h (28000 mph), and the Great Red Spot in its atmosphere is the **largest known hurricane** in the Solar System.

GAS GIANT Jupiter and its moons enabled Galileo to prove that the Earth goes round the Sun.

Jupiter's moons

Jupiter has four large moons (up to 5000 km/3000 miles in diameter) and at least 12 small ones (10-260 km/6-160 miles in diameter). The large moons – Callisto, Europa, Ganymede and Io – were discovered by Galileo in 1610, and are known as Galilean moons. Ganymede is the largest, Europa is the smallest, Callisto is heavily cratered, and Io is the **most volcanically active** place in the Solar System.

Saturn ❼

Named after the Roman god of time, it takes this gas giant **29 years to orbit the Sun**. Saturn is 95 times heavier than Earth. Its three main rings, and four fainter rings, are made of ice particles. The outermost rings are 275000 km (170000 miles) in diameter. Saturn has **30 moons**.

SATURN Despite its great size, its density is so low that it would float on water.

Uranus ❽

Uranus was the first planet to be discovered since classical times – by Sir William Herschel in 1781. It is named after Saturn's father. Uranus **rotates on its side**. Summers and winters are 21 Earth years long. Its surface temperature is -220°C (-364°F). Uranus has 11 faint narrow rings, first noted in 1977. It has **21 known moons**, some discovered very recently.

URANUS Its blue colour is due to its methane-rich atmosphere.

Neptune ❾

Discovered in 1846 and named after the Roman god of the sea. Like Uranus, Neptune appears blue due to methane in its atmosphere which absorbs red light. It has the **strongest winds** in the Solar System (2000 km/h/1240 mph). A Great Dark Spot of storms used to be visible in its atmosphere but this has now disappeared. Neptune has four faint rings and eight moons, named after characters from Greek myths. Its moon Triton is the **coldest place** in the Solar System (-236°C/-392°F).

NEPTUNE High-speed 'jet streams' in the upper atmosphere show dark blue around the planet's equator.

Pluto ❿

Discovered in 1930 by Clyde W. Tombaugh, and named after the Roman god of the underworld by an 11-year-old schoolgirl, Pluto is the **smallest** planet, smaller even than Earth's moon. Despite some debate, Pluto's planet status is definitely confirmed. Pluto's **unique oval orbit** sometimes brings it closer to the Sun than Neptune. Pluto has one moon, Charon, that is just over half its size.

NEIGHBOURS Pluto and Charon photographed by the Hubble telescope.

The Moon

Core facts ❶

◆ The Moon is around **4.5 billion years old**, which places its formation very close to that of the Earth.

◆ It was **probably formed** as the result of a large asteroid or small planet colliding with the young Earth, throwing rocky debris into space that gradually coalesced.

◆ The Moon **orbits the Earth**, from east to west, at a distance of 384000km (240000 miles). It is held in orbit by Earth's gravity.

◆ The Moon takes the same time to spin on its axis as it does to orbit the Earth (27–29 days). So the same side always faces the Earth.

◆ Though appearing to shine, the Moon is actually a **dark object** that reflects light from the Sun – it does not radiate its own light.

The visible face of the Moon

The visible side is heavily cratered from asteroid and meteorite impacts that occurred during the early Solar System. Craters measure up to 240km (150 miles) across. The areas that appear lighter from Earth are highlands. Rocks from these areas are dated at over 4 billion years. The dark patches, known as seas or maria, are low-lying areas that were once flooded by lava. Rocks here are 3–3.9 billion years old.

MOON FEATURES Many craters and maria can be seen from Earth with the naked eye.

Mare Frigoris
Plato
Posidonius
Mare Serenitas
Mare Imbrium
Mare Tranquillitatis
Aristarchus
Copernicus
Langrenus
Oceanus Procellarum
Kepler
Theophilus
Cyrillus
Mare Nectaris
Ptolemeus
Catharina
Alphonsus
Grimaldi
Piccolomini
Gassendi
Mare Nubium
Tycho
Clavius

★ 790

Dark side

The **first photograph** of the Moon's far side was taken in 1959 by Soviet probe Luna 3. It has fewer 'seas' of lava and more mountains than the side we see. 'Dark side' is a misnomer: the Moon orbits the Sun, so at some point sunlight falls on its whole surface, but we only ever see light on the side facing Earth.

THE FAR SIDE The crust is thicker and more pitted.

Mythology ❸

In Greek/Roman mythology several deities were associated with the Moon. The Greek deities were **Artemis**, goddess of chastity, the Moon and the hunt; **Hecate**, goddess of the Moon and the underworld; and **Selene**. **Diana** was the Roman equivalent of Artemis. **Luna**, another Roman goddess of the Moon, was depicted as a charioteer.

What is the Moon made of? ❻

The Moon has a **layered structure** similar to Earth, and is made of the same chemical elements, although in different proportions. Moon rocks include lavas from outpourings caused by meteorite impacts; calcium-rich rocks from the Moon's early formation; and breccias, fragments of rock cemented together. **Surface ice** was recently discovered in the dark corners of craters near the Moon's south pole.

WEIRD AND WONDERFUL ❼

The bright light of the full moon nearest the autumn equinox in late September gave Northern Hemisphere farmers and their labourers extra time to bring in the harvest – hence its name: the **harvest moon**.

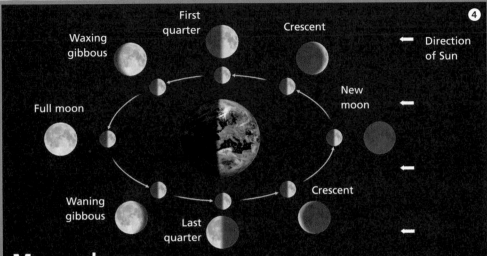

Moon phases

As the Moon orbits the Earth, the amount of its lit surface that is visible from Earth changes. These phases of the Moon begin with the **new Moon**, when only a tiny sliver is visible. The lit area increases (waxes) until we see the whole or **full Moon**, then decreases (wanes), ending with three days when the Moon is totally dark.

The inner circle of the diagram shows the Moon's position relative to the Sun and Earth. The outer circle shows what is visible from Earth in each phase.

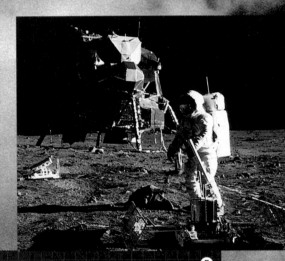

APOLLO 11 Buzz Aldrin sets up a seismometer on the Moon. ❺

Important moon missions

Date	Mission	Crew	Events
31 Jan 66	Luna 9 (USSR)	Unmanned	First soft landing.
21–27 Dec 68	Apollo 8 (USA)	Borman, Lovell, Anders	Orbit of Moon and view of the far side.
18–26 May 69	Apollo 10 (USA)	Stafford, Young, Cernan	Tested LM; took orbiter to 14 300 m (47 000 ft) above surface.
20 July 69	Apollo 11 (USA)	Armstrong*, Aldrin*, Collins	First manned landing, in Sea of Tranquillity.
19 Nov 69	Apollo 12 (USA)	Conrad*, Bean*, Gordon	Pinpoint landing of LM 183 m (600 ft) from Surveyor 3.
11 April 70	Apollo 13 (USA)	Lovell, Swigert, Haise	No landing, mission aborted.
10 Nov 70	Luna 17 (USSR)	Unmanned	Landed rover that sent TV pictures of surface.
5 Feb 71	Apollo 14 (USA)	Shepard*, Mitchell*, Roosa	Shepard played golf.
30 July 71	Apollo 15 (USA)	Scott*, Irwin*, Worden	First use of Lunar Roving Vehicle (moon-buggy).
20 Apr 72	Apollo 16 (USA)	Young*, Duke*, Mattingly	Young later first commander of Space Shuttle.
11 Dec 72	Apollo 17 (USA)	Cernan*, Schmitt*, Evans	Longest Apollo mission (301h, 52m), last manned visit to Moon.
6 Jan 98	Lunar Prospector	Unmanned.	Suggested presence of water in polar regions.

* The 12 men who have walked on the Moon.

The night sky

Core facts

◆ Constellations are **star patterns** which are visible in the night sky. Other bright night-sky objects include planets, nebulae, galaxies and supernovae.
◆ Forty-eight constellations in the Northern Hemisphere were listed by **Ptolemy** in the 2nd century AD and many were first recorded in Babylonian times.

◆ **Eighty-eight** constellations are currently identified. The constellation name refers to the **whole area of sky** occupied by a constellation, as well as to the star pattern.
◆ **Different** star patterns are seen in the Northern and Southern Hemispheres.
◆ Northern Hemisphere constellations known today were listed by **Helvelius** in 1687.

The moving sky

Since the Ancient Greeks, astronomers have called the night sky the **celestial sphere** because, seen from Earth, it looks like a transparent sphere with the stars attached to it. Earth's Equator and poles are projected onto the sphere to give the celestial equator and celestial north and south poles, which are used to map the position of stars in the sky. Because the Earth is spinning, the constellations appear to move across the sky. The portion of the celestial sphere that can be seen by the observer depends on their position on the Earth and the Earth's position in its orbit. An observer always has part of the sphere blocked out by the Earth itself. Although constellations look flat, the stars in them vary in distance from Earth. Stars Betelgeuse and Mintaka in Orion, for example, are 330 and 2300 light years away. Over time, constellations change shape because individual stars change position in relation to each other.

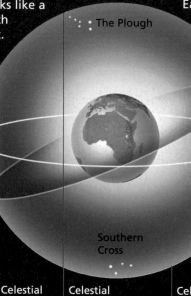

The Plough

Southern Cross

Celestial equator | Celestial north pole | Celestial south pole

CELESTIAL SPHERE The imagined transparent sphere around the Earth onto which the stars and constellations are projected.

Constellations ❸

The majority of **Northern Hemisphere** constellations that we recognise today were named by the ancient Greeks after characters or creatures in mythology. **Southern Hemisphere** constellations were mapped and named in the 16th–18th centuries, mainly by the Dutch, as European seafarers explored the southern oceans. A number are named after scientific instruments, including Antlia (the air pump), Telescopium (the telescope) and Horologium (the pendulum clock).

ORION The Hunter is one of the few constellations visible in both hemispheres.

CRUX More commonly known as the Southern Cross, Crux is an important aid to navigation.

URSA MAJOR Better known as the Great Bear, this constellation contains the Plough.

SCORPIUS The Scorpion contains the sky's brightest X-ray source. It is also a zodiac constellation.

Constellation records and facts ❹

◆ Hydra, the Water Snake is the **largest** constellation.
◆ Crux, the Southern Cross, is the smallest. It appears on the flags of Australia, New Zealand, Brazil and Western Samoa.
◆ Only a few constellations are **visible in both hemispheres** – for instance Orion and Aquila, the Eagle. Orion's belt lies on the celestial equator.
◆ **Pegasus** is visible in the Northern Hemisphere in autumn upside down and in the Southern Hemisphere in spring the right way up.
◆ The **12 zodiacal constellations** are intersected by the **ecliptic** (the plane of the Solar System). This means that, seen from Earth, the Sun, Moon and all the planets apart from Pluto travel through them. Gravitational effects on Earth since ancient times mean that the ecliptic now passes through a 13th constellation, Ophiuchus, the Serpent Holder.

The Pole Star

The star closest to the north celestial pole is known as the Pole Star. Currently this is a star called **Polaris**, in the constellation Ursa Minor. Whereas other stars move across the night sky, due to Earth's spinning, Polaris remains in the same position. There is no single star marking the south celestial pole, but the Southern Cross passes through it at times.

WEIRD AND WONDERFUL ❼

Stars **twinkle** because light from them travels through the Earth's atmosphere. The different layers of the atmosphere are different temperatures and bend the light, making the star appear to be moving slightly.

Other sights in the night sky ❺

Around 2000 stars can be seen with the naked eye in the night sky, but the **brightest bodies** are actually planets – Venus, Mercury, Mars and Jupiter. Of these Venus, also called the Morning (or Evening) Star, is the brightest. **Alpha Centauri**, one of the nearest stars to Earth, can only be seen in the Southern Hemisphere. The pale band across the sky is the **Milky Way Galaxy**, seen from the side. The **Andromeda Galaxy**, over 2.2 million light years away, is also visible. Other night-sky spectacles include supernovae – transient exploding stars – and nebulae, sites of star birth, such as the Crab Nebula.

STARFIELD The two bright stars above the mountain to the left are part of the Plough. They point towards Polaris on the far right.

Ursa Major ❻

The constellation Ursa Major in the Northern Hemisphere literally means '**Great Bear**' after the legend of Callisto, a nymph who was turned into a she-bear by Zeus and placed in the heavens. The seven brightest stars in the constellation are variously called 'The Plough', 'The Big Dipper' and 'The Wagon', after their distinctive shapes in the night sky.

Comets and meteorites

Core facts ❶

◆ **Comets** are bodies of ice and dust. In orbit in the outer reaches of the Solar System they are dark and cold, but when they approach the Sun they heat up and release glowing tails of dust and gas blowing away from the Sun.
◆ Fragments of rock up to 1 m (3 ft) wide are called **meteoroids**. When they enter Earth's atmosphere they burn up, producing showers of light called **meteors** or 'shooting stars'.
◆ Fragments and particles of rock that reach Earth are known as **meteorites**.
◆ **Asteroids** are rocky bodies that orbit the Sun mainly between Mars and Jupiter.
◆ By convention, **asteroids are named** by their discoverer, **comets** after their discoverer, and **meteorites** after their landing site.

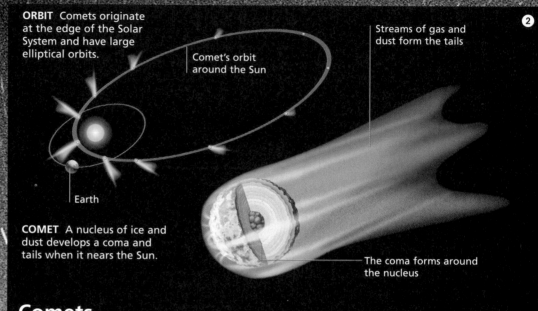

❷

ORBIT Comets originate at the edge of the Solar System and have large elliptical orbits.

Comet's orbit around the Sun

Streams of gas and dust form the tails

Earth

COMET A nucleus of ice and dust develops a coma and tails when it nears the Sun.

The coma forms around the nucleus

Comets

Comets have three components: a nucleus, coma and tails. The **nucleus** is a solid core of ice and 'dust' (fragments of rock from sand to boulder-sized). Over 80 per cent of the ice is water, the rest is frozen gases. The **coma** is a kind of atmosphere of dust and gases around the nucleus that burns off when it heats up as it approaches the Sun. Comets have **two tails**, dust and gas, which form when dust and gas particles are pushed away from the coma by solar wind and radiation. Gas tails can reach 100 million km long (60 million miles) and always point away from the Sun.

Famous comets ❸

Comets' orbits are divided into two types: short-period, which take up to 200 years to orbit the Sun; and long-period, which take more than 200 years. Short-period comets originate in the **Kuiper belt** – debris orbiting beyond Neptune and Pluto. Long-period ones originate in the **Oort cloud**, orbiting the Sun far beyond the planets. Comets can be dislodged from their orbits by the gravity of a star and pulled into the inner Solar System.

Name	Date discovered	Return period	Last seen
Halley	239 BC	76 years	1986
Encke	1786	3.3 years	2002
Swift Tuttle	1872	125 years	1996
Hale-Bopp	1995	18000 years	1997
Hyakutake	1996	14000 years	1996
Shoemaker-Levy	1993	Crashed into Jupiter	1994

WEIRD AND WONDERFUL ❹

Meteorites give scientists a unique opportunity to study fragments of the Solar System. A meteorite from Mars, which was found in Antarctica in 1984, bore shapes resembling microscopic fossils.

Collision course

Although most asteroids pose no threat to Earth, around 2000 large asteroids (measuring 1 km/$\frac{1}{2}$ mile or more) closely approach or cross Earth's orbit. They are officially known as NEOs: Near-Earth Objects.

METEOR CRATER Arizona, USA, 1265 m (4150 ft) wide.

Meteorite impact structures

There are about 150 confirmed meteorite impact structures on Earth. Meteoroids rarely reach the surface, normally burning up in the Earth's atmosphere.

Crater	Location	Diameter km (miles)	Age (million yrs)
Vredefort	South Africa	250 (155)	1970
Lake Acraman	Sourh Australia	160 (100)	600
Sudbury	Ontario, Canada	140 (87)	1840
Popigai	Russia	100 (60)	39
Lake Manicouagan	Quebec, Canada	70 (43)	210
Kara	Russia	50 (30)	57
Manson	Iowa, USA	35 (22)	70
Steen River	Alberta, Canada	25 (15.5)	95
Ries	Germany	24 (15)	15

Notable asteroids

Name	Claim to fame
Ceres, Pallas, Vesta, Hygiea	Known as 'The Big Four'
Ceres	Largest known – 940 km (580 miles) wide – and the first to be discovered (in 1801), by the Italian astronomer Giuseppi Piazzi
Spock	Named after the *Star Trek* character
Ida	Has its own moon (Dactyl)
323 Brucia	First to be captured by photography (1891)
Gaspra	First photographed by probe (1991, Galileo)

Meteorite origins

The vast majority of the **22 000** meteorites found on Earth are debris from the asteroid belt. Only 18 meteorites are thought to derive from the Moon, and only 14 from Mars. Some meteorites may have originated from comets.

Meteorites are most likely to be found in large open spaces such as ice fields and deserts, where they would not have been buried by sedimentation or rock folding, covered by vegetation or built upon. Over 17 000 meteorite samples have been collected in Antarctica alone.

METEOR PIECE A fragment of iron meteorite from Meteor Crater, Arizona. Meteorites can be made of metal, rock or a combination of the two.

Meteor showers

Most meteor showers occur when Earth passes close to a trail of dust from a comet.

Name	Dates visible
Quadrantids	1–6 January
Lyrids	19–25 April
Perseids	27 July–17 August
Taurids	25 October–25 November
Leonids	14–20 November
Geminids	8–14 December
Ursids	19–24 December

SHOWER OF LIGHT Meteors from the Leonids shower that occurs each November are seen as bright straight streaks against the night sky. The curved lines are star trails.

Planet Earth

Core facts ❶

◆ Earth **is unique** in the Solar System in being the only planet known to support life and to have liquid water at its surface.
◆ Earth **formed**, along with the rest of the Solar System, around 4.6 billion years ago.
◆ Earth's **physical environments** can be divided into the **solid earth**, the **atmosphere** made up of layers of gas, and all the water on the planet, known as the **hydrosphere**.
◆ Earth's **gravity** influenced the formation of the planet, holds its atmosphere in place, and keeps the Moon in orbit around it.
◆ **Eclipses** occur when the Sun, Moon and Earth are directly in line.

1 4.6 BILLION YEARS AGO The Earth and the other planets were created from a mass of gas and colliding dust particles that swirled around the young proto-Sun, being pulled together by gravity.

Stages of formation ❷

It took billions of years for the Earth to evolve from a mass of spinning gas and dust into a life-supporting planet. The Earth is a **dynamic system**. Its shifting surface plates mean there is a constant recycling of elements as rocks are created and destroyed at plate margins. Earth's three **physical environments** – atmosphere, hydrosphere and solid earth – are in perpetual flux and interaction. The biosphere, Earth's living skin, is also in constant interaction with these three environments.

2 4.6–4.2 BILLION YEARS AGO The Earth heated up from the decay of radioactive elements and became molten. Iron sank to form the core, while lighter elements formed the outer Earth.

3 4.2–3.8 BILLION YEARS AGO The hot surface cooled and hardened. Meteorite bombardments and volcanic eruptions released water vapour from Earth's interior, and this condensed into surface water. The first primitive bacterial life-forms emerged in the primeval seas.

Shape and dimension ❸

The Earth is 12 756 km (7926 miles) in diameter at the Equator. It is not a perfect sphere. The Earth **spins** at 1600 km/h (1000 mph) at the Equator but more slowly at the poles. The centrifugal force caused by this variation causes the planet to flatten at the poles and bulge outwards at the Equator.

SPINNING EARTH It takes the Earth 23 hours and 56 minutes to spin once on its axis.

Earth's orbit and spin ④

Earth is held in orbit by the gravitational field of the Sun. It takes 365.24219 days to orbit once. As it orbits, Earth rotates on its axis, creating day and night on the sides facing towards and away from the Sun respectively. One rotation takes just under 24 hours.

Earth's axis tilts at an angle of 23.5° to the plane of rotation. This angle varies from 21.5° to 24.5° over a period of 41 000 years. The axis also wobbles, like that of a spinning top, an effect known as precession.

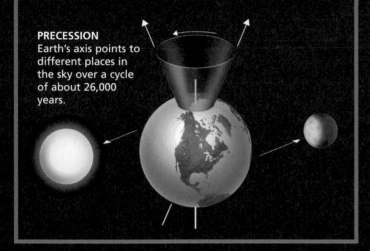

PRECESSION
Earth's axis points to different places in the sky over a cycle of about 26,000 years.

4 3.8–2.5 BILLION YEARS AGO
Photosynthesising bacteria pump oxygen into the atmosphere. Weather systems establish. Geological processes act on the early continents.

5 2.5 BILLION–6 MILLION YEARS AGO
The atmosphere becomes oxygen rich and the ozone layer forms. Complex life evolves in the oceans, plants colonise the land, followed by insects, amphibians, reptiles and mammals. Man's ancestors evolve.

Earthbound

Gravity is the attractive force that a large body, such as the Earth, exerts on other objects. The larger the body, the greater the force. The Earth is 81 times more massive than the Moon, and objects weigh six times more on Earth than on the Moon.

Marking out the surface ⑤

The **Equator** (0 degrees latitude) encircles the globe at its greatest diameter and divides the Earth into Northern and Southern Hemispheres. The Tropic of Cancer lies between the Equator and North Pole and the Tropic of Capricorn between the Equator and South Pole. Lines of **latitude** are horizontal. Lines of **longitude** are vertical and meet at the poles (the map is a Mercator projection).

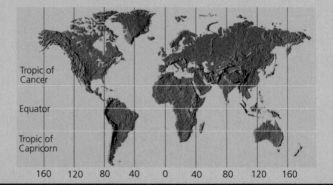

Tropic of Cancer

Equator

Tropic of Capricorn

160 120 80 40 0 40 80 120 160

Shadows in space ⑥

◆ An **eclipse** occurs when one object in space passes through the shadow of another. The term usually refers to eclipses of the Sun and Moon. Eclipses can be partial, or total – completely blocking the body from view.

◆ A **lunar eclipse** occurs when the Moon passes through the shadow cast by Earth on the side away from the Sun.

◆ A **solar eclipse** happens when the Moon passes between the Earth and the Sun.

LUNAR ECLIPSE The Earth's shadow plunges the Moon into total darkness.

SOLAR ECLIPSE The Moon blocks out the Sun as viewed from specific points on the Earth's surface.

Earth's structure

Core facts ❶

◆ As the Earth formed, gravity drew the **heavier elements** such as iron towards the centre, while buoyant masses of **rock** rose towards the surface and solidified.
◆ Earth has three main layers: the **crust**, **mantle** and **core**.
◆ The surface layer is split into **plates**, which move around on the underlying layer. The continents and oceans sit on these plates.
◆ At the centre is the **core**, where Earth's magnetic field is generated. The **magnetic field** extends for more than 60000km (37000 miles) into space, where it provides protection against radiation from the Sun.

Inside the Earth ❷

◆ **The crust** forms a solid, rocky outer skin. Beneath the oceans it is approximately 7km (4.5 miles) thick. Beneath the continents it is 35–40km (22–25 miles) thick.
◆ **The mantle** is 2900km (1800 miles) thick, and makes up more than 82 per cent of Earth's volume. It consists of molten rock called magma.
◆ **The core** divides into a hot, liquid, iron-rich outer and a solid inner core.
◆ The crust and top layer of the mantle form a cool, rigid layer, the **lithosphere**. The upper mantle forms a heat-softened layer, the **asthenosphere**.

EARTH LAYERS The mantle (purple) rests on the inner (red) and outer (orange) core.

On the move ❸

The rigid **lithosphere** is broken into slabs, or plates, which move around on the mantle below at an average rate of about 5cm (2in) each year.

The grinding and pushing of plates creates mountains and volcanic and earthquake activity along their boundaries. Where plates are moving apart, faults, rifts and seafloor spreading occur.

There are seven major plates – the North American, South American, Pacific, African, Eurasian, Australian and Antarctic – plus several smaller plates.

RIFT VISIBLE A gash in the landscape of Iceland marks the meeting place of the North American and Eurasian plates.

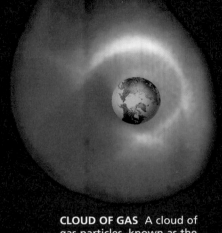

CLOUD OF GAS A cloud of gas particles, known as the magnetosphere, is trapped by Earth's magnetic field.

Magnetic attraction

④

The Earth's **magnetic field** consists of invisible lines of magnetic force that are generated in the liquid outer core. These emerge at the magnetic south pole, located near McMurdo Sound in Antarctica, and re-enter at the magnetic north pole, near Prince of Wales Island in the Canadian Arctic.

The **magnetic poles** closely coincide with but do not match the geographic poles, and their positions change over time. The magnetic north pole is currently drifting westwards at a rate of 0.2° a year.

Every half million years or so, Earth's magnetic field flips over. The process can take 1000–1500 years, during which the magnetic field weakens and the poles move around before settling in reverse positions and building up strength again.

WEIRD AND WONDERFUL

⑤

The **magnetosphere** stretches away in a long tail on Earth's night side. From time to time this catapults back towards Earth and swirls around the planet, causing magnetic disturbances within Earth's magnetic field.

★ 800

Van Allen belts

The region enclosed by Earth's magnetic field is called the **magnetosphere**. It captures harmful high-energy charged particles from the Sun and traps them in two belts, named after their discoverer James van Allen, that encircle the Earth above the Equator.

RADIATION BELTS The belts circle the Earth one inside the other.

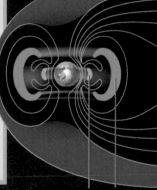

Van Allen belts

Outer edge of magnetosphere

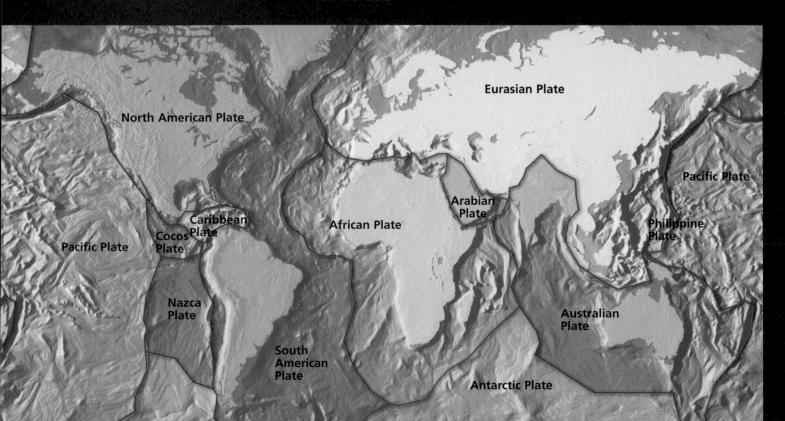

Eurasian Plate

North American Plate

Pacific Plate

Caribbean Plate

Arabian Plate

Cocos Plate

African Plate

Philippine Plate

Pacific Plate

Nazca Plate

Australian Plate

South American Plate

Antarctic Plate

Rock types and formations

Core facts ❶

◆ **Rocks are created** by heat and pressure within Earth's crust. Over time, rock alters its form, with old rock being recycled into new.
◆ The **three main types** of rock are: sedimentary, igneous and metamorphic.
◆ **Igneous** rock makes up most of the solid part of the planet, partially covered with thin layers of sedimentary and metamorphic rock.
◆ **Sedimentary** rock accounts for only 5 per cent of rock in the outer 16km (10 miles), but 75 per cent of all rock visible at the surface.
◆ Rocks provide an almost unlimited source of materials for buildings, roads and a range of industrial processes.

Continuous production

Igneous rock (igne=fire, ous=full of) forms when molten rock, or magma, migrates up to the crust, where it cools and solidifies. Common igneous rocks include peridotite, basalt, gabbro, dolorite and granite. Metals such as gold, silver and copper form around igneous granite.

Igneous rocks exposed at the surface undergo weathering and erosion, which produces sediment. This is eventually carried out to sea, where it collects in layers. These become increasingly deeply buried and compressed by the weight of overlying layers to form **sedimentary rock**. Examples include sandstone, limestone, chalk and shale. Sedimentary rock is the principal source of coal, oil and gas, and metals such as iron and aluminium. It provides cement and other natural building materials. Most of the underground fresh water on the planet is stored in it. Some sedimentary rock contains the fossil remains of past life. Also, because sedimentary rock is made up of the remains of earlier rock, it provides clues to the Earth's geological history.

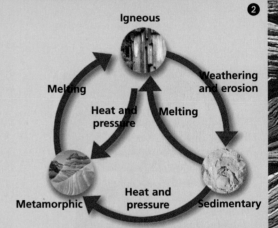

❷

Igneous

Weathering and erosion

Melting

Melting

Heat and pressure

Metamorphic

Heat and pressure

Sedimentary

Metamorphic rock (meta=change, morph= form) is produced from pre-existing rock that is altered, but not melted, by heat and pressure deep below the Earth's surface. Common metamorphic rocks include marble, which forms from limestone, and slate from shale or mudstone. If the pressure and heat increase to the point where the rock melts, magma forms to produce new igneous rock.

Skeletal rock ❸

Limestone represents 10 per cent of all sedimentary rock. Most limestone forms in marine environments from the fossilised remains of the shells and skeletons of tiny organisms, which rain down on the ocean floor and gradually become compacted into layers of sediment.

There are two types of limestone in which its biological origins are very obvious: **coquina**, a coarse rock composed of poorly cemented shell fragments; and **chalk**, a porous, soft rock consisting almost entirely of the shells and skeletons of marine creatures.

★ 285

Basalt pillars

Basalt is a particularly hot, fluid lava that spreads easily and can cover large areas. As the lava slowly cools it contracts and cracks in a regular five or seven-sided pattern. The cracks extend from top to bottom of the rock, resulting in a network of pillars or columns with very thin gaps between them. **The Giant's Causeway** on the Antrim coast of Northern Ireland and **Fingal's Cave** on the Hebridean island of Staffa are good examples.

ORGAN PIPES Straight-sided columns are clearly visible in these basalt pillars at Sawn Rocks, Australia.

WEIRD AND WONDERFUL

4

The **Rock of Gibraltar** is the remnant of a natural dam that cut off the Mediterranean from the Atlantic around 8 million years ago as the African and Eurasian plates collided and forced the seafloor upwards.

Common uses of different rocks **5**

Rock	Characteristics and uses
Granite	Resistant to weathering. Often at core of eroded mountains. Building material, paving, tombstones.
Pumice	Volcanic origin and glassy texture. An abrasive in cleaning, polishing and scouring; an aggregate in masonry, concrete, insulation and acoustic tiles.
Chalk	Varied uses in fertiliser, cement, putty, as whiting in ceramics, in rubber, paper, paints, crayons, plastics, cosmetics (but not school chalk).
Clay	Used in tiles, bricks, ceramics, sculpture, paper coating, water softener, Portland cement.
Conglomerate	Made up of gravels. Used for construction and in road-building.
Flint	Used in building, road construction, sandpaper, as a grinding agent and in concrete aggregate.
Limestone	Used in flooring, Portland cement, fertiliser. An ingredient in steel and glass.
Sandstone	Varieties include quartz, sandstone, arkose and graywacke. Used as a building material.
Shale	Most common sedimentary rock. Crumbles easily. Ingredient in tiles, bricks, pottery, ceramics, cement.
Slate	Breaks into flat slabs. Used for electrical panels, tabletops (snooker, laboratories), blackboards, roofing and flooring.
Marble	Relatively soft and easy to cut and shape. Used for building, flooring, sculpture, tabletops.

UNDULATING LAYERS These sandstone hills in Arizona, USA, show how layers of sedimentary rock become distorted by stresses within Earth's crust and twist out of shape.

THE LINCOLN MEMORIAL This famous statue of Lincoln is carved in marble.

Weathering

Core facts ❶

◆ **Weathering** is the gradual breaking down of rock exposed at the Earth's surface. Joints and fissures in rock let in air and water and facilitate the process.
◆ **Physical forces** such as ice and tree roots break up large expanses of rock into smaller pieces, and expose the surface to attack from acids dissolved in water and from oxygen in the air.
◆ Water and oxygen react with veins of minerals in rock to produce areas of weakness.
◆ Weathering exposes deposits of ores and minerals in rock, and creates the sediment that forms soil.

Physical forces ❷

Physical weathering includes the action of ice and frost and the effects of plants and animals.

When water gets into cracks in rock, it freezes and expands. This expansion can exert a force up to $635\,kg/cm^2$ (1400 lb/sq in). With repeated **freezing** and **thawing** cracks gradually enlarge to the point at which pieces of rock break off. The process, known as **ice-** or **frost-wedging**, is most pronounced on mountains, where the daily cycle of freezing and thawing, even in summer, creates large accumulations of rock rubble, called **talus** or **scree**, around the base of steep slopes.

Plant and tree roots get into surface cracks in rock, where they can exert a force so strong that it widens a crack, opening it up to **chemical weathering**.

Large animals contribute by stepping on or crushing stones and pebbles. Even earthworms help to break up rock fragments.

THE POWER OF ICE
Ice-wedging is particularly active in mountains, where meltwater from snow seeps into cracks then freezes and expands at night.

SLOW DISINTEGRATION
In Bryce Canyon, USA, weathering has accentuated the differences in rock types, and landslides and erosion have removed the resulting debris, leaving an exotic array of rock shapes.

Exfoliation domes ❸

When an area of **igneous** rock is uncovered by the erosion of overlying rock layers, the pressure on it is reduced and the rock expands upwards. The outer layer of granite expands more than the rock underneath and slabs break off, forming an **exfoliation dome**. Stone Mountain in Georgia and El Capitán in Yosemite were formed this way. A combination of **exfoliation** and **erosion** by glaciers produced Half Dome in Yosemite.

YOSEMITE Half Dome rises more than 1200 m (4000 ft) above the valley floor.

Chemical forces ❹

Chemical weathering occurs when water or air come into contact with rock surfaces. For example, **acids** dissolved in water attack soluble rock such as limestone, while **oxidation** works on iron-rich rocks such as basalt. These processes change the mineral content in rock into other, often softer, minerals, many of which are soluble in water and get washed away. Most weathering happens within a few metres of the Earth's surface. Heat speeds up the rate of chemical reactions.

GRANITE BOULDERS Edges and corners weather faster than flat surfaces, and all boulders gradually develop a rounded shape, like these in the Namib desert.

★ 480

Onion-skin weathering

In arid areas where rock surfaces are exposed to high daytime temperatures and low night-time ones, the surface layers expand in the day and contract at night. Eventually, the outermost layer separates from the rock beneath and breaks off, in a process known as onion-skin weathering. The **Devil's Marbles** in Australia were formed by this process.

Making soil ❺

Much of the Earth's surface is covered with a blanket of disintegrated and decomposed rock and mineral fragments produced by weathering. Where this is mixed with water, air and decayed plant and animal remains (humus), soil forms. The mineral content of the bedrock determines the nutrient richness of the soil produced.

The soil-forming process works from the surface downwards, and variations in composition and texture evolve at different depths. A vertical section through the layers is called a soil profile.

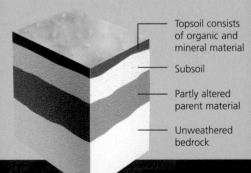

- Topsoil consists of organic and mineral material
- Subsoil
- Partly altered parent material
- Unweathered bedrock

Tie-breaker ❻

Q: What common metal gives many sandstones their red colour?
A: Iron. In fact, it is rust (iron oxide) that makes the rock red. Oxygen in water combines with iron in rock to form a red rust, known as hematite, or a yellow rust, called limonite.

Erosion

Core facts ❶

◆ **Erosion is** the process of removing sediment and transporting it from one place to another. The main agents are water, wind, glaciers (see pages 92-3) and gravity (see pages 120-1).
◆ Rivers are the **most powerful** force of erosion, creating and deepening valleys through abrasion and dissolution.

◆ The **varying resistance** of different rocks to erosion produces tablelands and rock pillars.
◆ The erosion agent 'selects' sediment grains of an appropriate size. Wind picks up small grains, whereas a flooded river or glacier picks up any size from tiny particles to boulders the size of small buildings.

River gorges ❷

Where a river's flow is fast and strong, it lifts away loose particles of sand and gravel from the riverbed and bank. These particles abrade the river floor as they are carried along, and have the power to cut down even through bedrock. The silt-laden **Colorado River** carved out the Grand Canyon in this way, producing a gorge 2 km (1 1/4 miles) wide and up to 1.6 km (1 mile) deep in places.

When a river crosses **soluble bedrock** such as limestone some of the rock may dissolve out. Every year, rivers remove an estimated 3.5 million tonnes of rock from the continents by dissolution and carry it out to sea. The **Niagara River** carries 60 tonnes of dissolved sediment over the Niagara Falls every minute.

BEEHIVES In the Bungle-Bungle mountains of Australia, streams have created 'beehive' rock shapes and a deeply incised river valley.

The power of rain ❸

The erosive power of rainwater gradually removes succeeding surface layers, leaving erosion-resistant areas of rock as isolated outcrops or island mountains.

Feature	Type	Location
The Mittens	Desert buttes	Monument Valley, USA
Delicate Arch	Sandstone arch	Arches National Park, USA
Devil's Tower	Basalt rock	Wyoming, USA
Tepuis	Island mountains	Venezuela
The Meteora	Sandstone towers	Greece
Uluru	Sandstone hill	Australia
Wave Rock	Water and wind	Australia
The Pinnacles	Limestone pillars	Australia

WEIRD AND WONDERFUL ❹
Earth pillars have a variety of nicknames, including 'little men' and 'earth mushrooms' in northern Italy, 'ladies with their hats on' (*demoiselles coiffées*) in the French Alps, and hoodoos in North America.

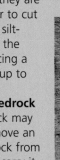

Window arch

⑤

Unlikely as it seems, water is responsible for the creation of rock arches, such as the one forming here. Rainwater washes away grains of sandstone, and in winter ice-wedging breaks off flakes and fragments, exposing new areas to erosion and gradually enlarging the hole into an arch.

Wind and sand

Wind erosion is only significant in **arid areas** and deserts, where it removes sand and soil. The **White Sands** region of New Mexico (USA) is covered by a layer of gypsum sand that has been eroded from mountains to the west and deposited here. As the sand is blown around, it is deposited around obstacles such as rocks and vegetation. The sand-laden wind scours away at these deposits, leaving trees with a hard casing of sand around them.

723

Earth pillars

In semi-arid areas loose surface material and soft, soluble rock can easily be washed away by sudden heavy rains, except where harder rocks protect the ground beneath them. One flash flood can wash away so much material that it leaves pillars of softer rock topped by erosion-resistant cap rocks.

PRECARIOUS BALANCE
A boulder sits atop a pillar of softer rock in the Painted Desert, Arizona.

Caves

Core facts

◆ **Caves are** natural underground cavities. The majority form in limestone regions, when water dissolves out and enlarges underground cavities in soluble limestone rock.

◆ The **calcium carbonate** dissolved out of the limestone and deposited on cave walls is called travertine, also known as dripstone.

◆ **Stalactites** and **stalagmites** are created by the endless dripping of water over large spans of time.

◆ **Non-limestone** caves include lava tubes, ice caves and sea caves.

◆ Caves have long been used by people and provide clues to human history.

Cave dwellers ❷

Many people worldwide carve out homes underground, often to escape the heat. They include the troglodyte communities of **Trôo** in central France, **Matera** in southern Italy, **Guadix** in southern Spain, **Matmata** in Tunisia, **Coober Pedy** in South Australia and parts of **Shaanxi** and **Shanxi** provinces in northern China.

DEER CAVE The largest cave passage in the world is in the Gunung Mulu National Park, Sarawak.

PALAEOLITHIC ART The paintings at Lascaux are around 15,000 years old.

Cave paintings ❸

Prehistoric cave paintings have been found at several locations in southern Europe.

◆ The **Lascaux** caves, discovered in 1940 in the Dordogne region of France, contain 600 painted or drawn animals and symbols, and a single human figure.

◆ **Les Trois Frères** cave in southern France contains paintings of human figures including the mysterious 'sorcerer' – half man, half stag.

◆ The **Font-de-Gaume** caves, also in the Dordogne, contain 200 painted or engraved figures, including bison, mammoth, reindeer, horse and auroch.

◆ At the **Altamira Cave** in Spain, discovered in 1868, the roof is covered in paintings, mostly of bison.

Records and famous finds ❺

Cave	Location	Details
Mammoth Cave System	Kentucky, USA	Longest cave system – 550 km (340 miles) of chambers and passages.
Krubera	Georgia	Deepest discovered – 1710 m (1 mile) deep. Discovered in 2002.
Sarawak Chamber	Gunung Mulu National Park, Sarawak	Largest single cave chamber – 700 m (2300 ft) long, 70 m (230 ft) high.
Cueva San Martín Infierno	Cuba	Tallest stalagmite – 67.2 m (220 ft).
Tham Sao Hin	Thailand	Tallest column – 61.5 m (202 ft).
Qumran	Dead Sea, Jordan	Discovery of Dead Sea Scrolls in 1947.
Lower Cave Chou-k'ou Tien	Peking, China	Archaeological site where 'Peking Man' – earliest fossils of *Homo erectus* – found.

FLUTED COLUMN The Grand Column in the Jenolan Caves, Australia, may be as much as one million years old.

Cave formation ❹

Caves form below the water table, in the wettest region of rock. Acidic groundwater follows lines of weakness in rock, enlarging them into passageways and caverns. If the water table drops, leaving the cavern relatively dry and exposed to air, stalactites and stalagmites form.

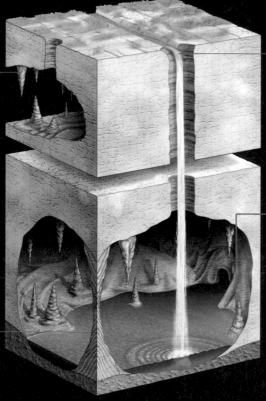

STALACTITE Stalactites (from the Greek *stalaktos*, 'oozing out in drips') hang from cave ceilings. Water oozes out of cracks in the ceiling and as it evaporates it leaves behind tiny deposits of calcium carbonate. Initially these create hollow tubes. Eventually the tube clogs up, and water runs down the outside, creating the typical stalactite shape.

COLUMN When a stalactite and a stalagmite meet and merge, they form a column.

POTHOLE A hole formed in a streambed.

STALAGMITE Stalagmites (from *stalagmos*, 'dripping') are deposits that build upwards on a cave floor from water dripping from the ceiling.

★ 856

Ice caves

Some limestone caves develop **permanent ice deposits**, including frozen lakes and ice draperies. The Eisriesenwelt system in the Austrian Alps is this type. It is the longest ice cave in the world, with 40 km (25 miles) of frozen landscape.

Ice caves also occur where streams hollow out cavities in glaciers.

CARVED OUT OF ICE An ice cave in Glacier Bay, Alaska.

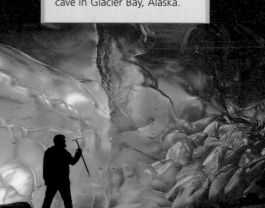

Fossils

Core facts ❶

◆ **The term** fossil comes from the Latin *fossilis*, meaning 'dug up'. Fossils are the remains of ancient organisms, as well as traces of their existence, preserved in rock.

◆ Most **fossils form** when the hard parts of animals, such as shells and skeletons, and less often plant material, become buried in sediment, and preserved as it converts into rock.

◆ Fossils of species that lived for just a short span of geological time, and are widespread around the world, allow geologists to date the rocks in which they are found to a reasonably specific time period. Such fossils are known as **index fossils**.

★ 404

Fossil jewels

Amber is fossilised resin produced by long-extinct coniferous trees. Most amber is around 70 million years old, and it was particularly fashionable in jewellery in the 19th century. The resin sometimes trapped small insects as it oozed out and solidified, resulting in detailed preservation. Black fossilised wood, known as jet, was also popular with the Victorians who used it polished, carved or faceted, especially as mourning jewellery.

CAUGHT IN TIME Amber preserves fossils complete with all their soft tissue.

Common types of fossil ❷

The most common fossils are of creatures that had some hard body parts, such as bones and shells.

Name	Form	Comments
Stromatolite	Colonial bacteria	Layered, mounded structures built up by colonies of bacteria trapping sediment. Up to 3500 million years old.
Ammonite	Mollusc with coiled shell	Similar in appearance to the modern Nautilus. Good index fossil. Sudden mass extinction around 65 million years ago.
Brachiopod	Shellfish with two valves	Once one of the commonest ancient life forms; also known as lamp shells as they resemble Roman oil-lamps.
Trilobite	Marine arthropod	These fossils are common because to grow trilobites had to shed their exoskeletons regularly, and these were often fossilised.
Belemnite	Marine mollusc with long, pointed conical shell	Related to modern cuttlefish, squid and octopus; once thought to be remains of a thunderbolt from heaven.
Crinoid	Echinoderm, sea lily	There are still 700 living species known.

AMMONITE This fossil was found at Lyme Regis, England.

TRILOBITE These creatures scurried around on the seafloor until their extinction 245 million years ago.

Fossil formation

The majority of fossils form as a cast or a mould in sedimentary rock.

A **mould** is an imprint of the organism's external form left after it has dissolved or decayed away. A **cast** forms if minerals dissolved in groundwater fill the mould or replace the organism. Shells, plant material and bones of vertebrate animals (including dinosaurs) that have been mineralised are said to have been **petrified** – literally turned into stone – when cavities and pores and hard parts are replaced by minerals such as quartz and calcite.

Carbonisation occurs when fine sediment encases an organism, and pressure gradually squeezes out all liquid and gaseous material, leaving a thin film of carbon in the rock. Leaves and insects are typically preserved this way.

The bodies of mammoths have been found frozen in permafrost, but this is not a common form of fossilisation.

BURIAL The empty shell of a dead marine animal is buried in sediment.

FOSSILISATION The sediment turns to rock, and the shell is mineralised.

EXPOSURE The rock is up-lifted and eroded, revealing the fossil.

Piltdown Man ❺

In 1910 geologist Charles Dawson found fragments of an ape-like cranium, jaw and teeth on Piltdown Common in Sussex, England. Two years later, Arthur Smith Woodward, keeper of the British Museum's palaeontology department, claimed that these fragments represented a missing evolutionary link between apes and humans.

Only in 1953-4 did tests reveal the find as a hoax: the skull was just 600 years old and the jaw and teeth belonged to an orang-utan.

IGUANADON Bipedal iguanadon was made four-legged, and its thumb spike erroneously placed on its nose, in this 19th-century reconstruction.

Misinterpretation of fossils ❹

Before geologists proposed that the Earth was millions, rather than thousands, of years old, various explanations for fossil remains were put forward.
◆ The Ancient Greeks thought the skulls of prehistoric elephants were the remains of Cyclops because the hole from which a trunk would have protruded resembled a giant eye socket.
◆ Marine reptiles such as icthyosaurs 'proved' the existence of 'sea dragons'. Icthyosaurs ate belemnites, whose bullet-shaped internal shells were thought to be thunderbolts sent from heaven.
◆ In the 18th-century, Johann Scheuchzer proposed that fossils were evidence of the Biblical Flood. He interpreted a fossil of a giant salamander as the remains of a man who had drowned.

Traces of life ❻

Fossils that are not actual remains of plants or animals but which provide evidence of the presence of prehistoric organisms are known as **trace fossils**, and include:
◆ **Tracks** and footprints made by animals in sediment that subsequently turned to rock.
◆ **Tunnels** made by creatures burrowing through sediment, rock or wood that later solidified or filled with mineral material.
◆ **Fossilised dung** and droppings, and dinosaurs' stomach stones.

ANSWERS

★ **65** A: Rift ★

70 C: Silurian ❸

84 True ❹

137 D: Carboniferous ❷ ❸

149 Towards each other ❹

172 Precambrian ❸

179 Laurasia and Gondwana ❸ ❹

288 Apart ❹

385 Cretaceous ❷ ❸

389 Pangaea ❸ ❹

410 C: Archbishop Ussher – of Armagh (1581-1656)

427 *Jurassic Park* – directed Steven Spielberg, 1993

531 Ice age ❸

597 Land bridge ❸

607 True ❹

989 The white cliffs of Dover ❷

1000 4.6 billion ❸

Earth's past

Core facts ❶

◆ The **sedimentary rocks** that make up 75 per cent of the Earth's land surface are composed of material derived from earlier rocks and from animal and plant remains. Sedimentary rocks provide clues to past Earth environments, such as seas and deserts, even though they might now be in areas of dry land or lush forest.

◆ This information, together with fossil finds and modern rock-dating techniques, has enabled scientists to piece together **Earth's geological history**.
◆ The **geological timescale** divides the 4.6 billion years of Earth's history into units of varying lengths of time.

Origins of names ❷

Several geological periods are named after the place where rocks of that age were first studied:
Jurassic after the Jura mountains.
Permian after the province of Perm in Russia.
Cambrian after the Roman name for Wales.

Others are named after a key characteristic of the rocks of the period, such as:
Carboniferous for its coal-bearing strata.
Cretaceous from the Latin for chalk (*creta*), after the extensive chalk deposits laid down at that time.

BOUNDARY LAYER A thin (roughly 2 cm/1 in) layer of clay appears in many places around the world between older Cretaceous rocks and younger Tertiary rocks. It is thought to be associated with a large meteorite impact that may also have caused the extinction of the dinosaurs.

Geological timescale ❸

Years ago	Era	Period	Epoch	Life	Key event
10 000-present	Cenozoic	**Quaternary**	Holocene		Post-glacial rise in sea levels
1.8 million-10 000	Cenozoic	**Tertiary**	Pleistocene		Last glaciation in progress. **Bering land bridge** links Asia with North America
5-1.8 million	Cenozoic	**Tertiary**	Pliocene		North and South America joined
23-5 million	Cenozoic	**Tertiary**	Miocene		Antarctic ice cap develops
34-23 million	Cenozoic	**Tertiary**	Oligocene		Himalayas begin to form. Red Sea opens up
57-34 million	Cenozoic	**Tertiary**	Eocene		Australasia separates from Antarctica. **Ice begins to form** at Poles
65-57 million	Cenozoic	**Tertiary**	Palaeocene		Rocky Mountains form. India begins to collide with Asia
144-65 million	Mesozoic	**Cretaceous**			Gondwana splits up
208-144 million	Mesozoic	**Jurassic**			Pangaea breaks into Laurasia in the north and Gondwana in the south. **Atlantic Ocean** forms
245-208 million	Mesozoic	**Triassic**			Deserts widespread
286-245 million	Palaeozoic	**Permian**			Supercontinent Pangaea intact
360-286 million	Palaeozoic	**Carboniferous**			Inland seas cover most of North America and Europe. Extensive coal-forming forests
408-360 million	Palaeozoic	**Devonian**			Mountain-building in north-west Europe and north-east America
438-408 million	Palaeozoic	**Silurian**			First coral reefs and **land plants**
505-438 million	Palaeozoic	**Ordovician**			Beginning of mountain-building in north-east America. Glaciation in the Sahara
550-505 million	Palaeozoic	**Cambrian**			Earliest evidence of complex life forms
4.6 billion-550 million	Precambrian				Birth of planet, formation of the crust, appearance of the **first life forms**, development of the atmosphere

Continents on the move ❹

The large plates into which **Earth's crust** is divided include whole continents and large areas of ocean, but no plate exactly matches the boundaries of a single continent. As the plates have moved, continents and ocean basins have changed shape, size and location.

Around **250 million years ago**, all the continents were joined into one supercontinent known as **Pangaea** (pan = all, gaea = Earth).

Around **200 million years ago** it began breaking up into smaller continents, which have gradually drifted to their present positions.

450 MILLION YEARS AGO Landmasses barely recognisable as today's continents were clustered in the Southern Hemisphere.

250 MILLION YEARS AGO The landmasses were joined in one vast supercontinent, Pangaea.

160 MILLION YEARS AGO Pangaea had split into two continents, Laurasia in the north – including Asia and North America – and Gondwana in the south, including South America, Africa, Australia and Antarctica.

65

Rifting apart

Weak spots in the Earth's crust allow magma to push upwards, warping and cracking the surface. As the two sides of the crack slowly pull apart, large slabs of rock sink, and the centre subsides to form a flat-bottomed trench. The East African Rift Valley, the Rhine Valley and the Rio Grande are all **continental rift valleys**.

The East African Rift Valley may represent the first stage in the break-up of a continent, much as happened with Pangaea.

PRESENT The continents are still drifting. Australia is moving north, Africa is likely to split along the Rift Valley and the Atlantic is widening.

DEVELOPING RIFT The East African Rift Valley may split Africa in two.

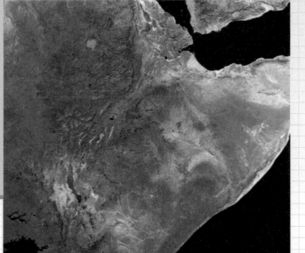

Clues to past connections ❺

Evidence that the continents were once joined in the massive supercontinent, Pangaea, includes:
◆ The fit of the **coastlines** of South America and Africa.
◆ **Mountain belts** that end on one coastline re-emerging on another across the ocean.
◆ Evidence of global **climate change** on different land masses.
◆ The same type of **fossil organisms** being found on several continents.

The numbers or star following the answers refer to information boxes on the right.

ANSWERS

5	**Blue** – was used as a dye for paints and fabric
61	**D: Paris** ❷
645	**C: White** ❷
646	**B: Silica** ❹
657	**Mineral** – from *The Pirates of Penzance*, 1879
755	**One** ❼
832	**C: Graphite** ❷ ❼
846	**Red** – it's an opaque type of quartz
906	**Asbestos** ❷
908	**Kryptonite** – radioactive remains of planet Krypton
940	**Sulphur** ❷
948	**Gunpowder** ❷
962	**Brine** – although has more salt than typical seawater
963	**Lot's wife** – fleeing from Sodom and Gomorrah
★ **966**	**Salary** ★
968	**Rock salt** ★
970	**Raise** – more heat energy is needed to make salted water boil

Minerals

Core facts ❶

◆ Minerals are **inorganic substances** that occur naturally in rocks. There are more than 3000 known minerals, but 99 per cent are made up of just eight elements: oxygen, silicon, aluminium, iron, calcium, sodium, potassium and manganese.

◆ Minerals occur in a pure state and in combinations, and include **metals and gems** as well as non-metallic minerals.

◆ The majority of mineral **deposits form** underground but some form on the seabed and others from the evaporation of mineral-rich waters at the Earth's surface.

◆ Minerals are **identified** by their colour, crystal structure, how they reflect light, their hardness, density, and the way that they break.

Common industrial minerals ❷

Mineral	Characteristics	Uses
Albite	White	Glass, ceramics
Asbestos	Whitish, fibrous	Fire-proof fibres
Corundum	Abrasive	Sandpaper, emery boards
Fluorite	Brittle, comes in several colours, including blue	Iron and steel-making, microscope lenses, pottery glazes and toothpaste. Blue fluorite is a decorative stone known as Blue John
Graphite	Soft, greasy	Pencil lead
Gypsum	White, chalky	Plaster-of-Paris, used for casts and moulds, dentistry, cement, classroom chalk
Halite	Colourless	Rock salt
Kaolinite	White clay	Medicine, chinaware, glossy paper
Phosphorus	Colourless, waxy	Fertilisers, fireworks, steel
Quartz	Hard, many varieties	Computer chip manufacture. Clocks and watches
Sulphur	Pale yellow	Paints, dyes, detergent, gunpowder, explosives, paper, medicines, insecticides, electrical insulation

EXPLOSIVE APPLICATION Sulphur is a versatile substance with many and varied uses. It is a vital ingredient in the manufacture of explosives, including fireworks.

How mineral deposits form ❸

◆ All rocks are composed of assemblages of minerals. The **heat and pressure** generated by geological processes, such as volcanism and metamorphism, can concentrate minerals into significant deposits.

◆ Some mineral deposits form on the surface when water rich in minerals **evaporates**.

◆ Minerals dissolved in seawater **crystallise** round volcanic vents on the seabed.

MAGMA CHAMBERS Heated fluids seep into the rock around igneous intrusions, and minerals crystallise out, forming veins.

EVAPORATION When water evaporates, salt, gypsum and other minerals build up around the water's edge.

SEA FLOOR Minerals crystallise round vents, or precipitate from sea water.

GLASS HOUSE ❹
Glass has become an important element in the construction of modern buildings, with some spectacular results.

Glass

The basic ingredient of glass is silica, which is derived from quartz sand. The silica is heated until it melts, at about 1100°C/2012°F, shaped and then rapidly cooled. Other important raw ingredients in glass manufacture are lime and soda. Borax is added to make glass **heat-resistant**. Lead oxide improves the **sparkle** and quality of cut glass. **Laminated** glass is made by sandwiching a layer of plastic in between two sheets of glass. This prevents pieces flying around if the glass is shattered.

Tie-breaker ❺

Q: What is the most common element in the Earth's crust?
A: Oxygen. By weight, 46.6 per cent of the crust is oxygen. It occurs as oxides, in combination with other elements. The next most common element is silicon, which accounts for 27.7 per cent of the crust.

WEIRD AND WONDERFUL ❻

The main source of **phosphorus**, used as fertiliser, is the mineral apatite, found in some sedimentary rocks, and also in deposits of the droppings (guano) of bats and birds.

★ 966

Salt

The preservative properties and dietary importance of salt have made it a valuable commodity since ancient times. Salt was included in offerings to the gods by the Romans and Greeks. Roman soldiers were paid an allowance of salt, a salarium (from which we get 'salary'). Salt was transported along the Via Salaria (Salt Route). Cakes of salt have been used as payment in Ethiopia and Tibet. Salt occurs naturally as halite, or rock salt, in some sedimentary rocks and it can be crystallised from evaporating sea water.

SALT-MAKING Salt crystallising out of sea water is swept into mounds.

Mohs' Hardness Scale ❼

A mineral's hardness is a measure of its resistance to scratching. It is established by scratching the mineral with a series of other minerals whose hardness is already known. Mohs' Scale provides a measure for comparison, from 1 (soft) to 10 (hard). It is linear up to 9, but diamond (10) is about ten times harder than corundum.

1	Talc
2	Gypsum
3	Calcite
4	Fluorite
5	Apatite
6	Orthoclase
7	Quartz
8	Topaz
9	Corundum
10	Diamond

The numbers or star following the answers refer to information boxes on the right.

ANSWERS

8	**Yellow** – elsewhere, more commonly called gold fever
10	**Lead white** – its use was banned in UK in the 1970s
33	**Silver** ❷ ❹
56	**Gold** ❷ ❹
213	**Silver** ❷ ❹
437	**Forty-niners** ❹
561	**Copper** ❸ ❹
565	**Gold** ❷
567	**Hallmark** ❻
611	**Platinum** ❹
612	**Carat** ❺
613	**Mercury** ❹
★ 614	**Fool's gold** ★
615	**Tin and lead** ❸
616	**Tungsten** – the metal with the highest melting point
617	**Aluminium** (found chiefly in the ore bauxite) ❶
618	**Nickel** ❹
619	**Panning for gold** ❹
620	**Fort Knox** – about 25 miles (40 km) from Louisville
815	***Full Metal Jacket*** – title refers to a kind of bullet
836	**A: Titanium** – first used by Lockheed in 1959
925	**Gold** – alchemists also sought eternal youth through the 'elixir of life'

Metals

Core facts ❶

◆ Most metals are spread widely through the **Earth's crust**. Aluminium is the most common, its compounds forming 8 per cent of the crust.
◆ Most major deposits formed around present and past **plate boundaries** (page 60-61), and where magma and hydrothermal solutions are constantly circulating through rocks. Weathering and erosion help to produce localised concentrations of useful metals and ores.

◆ A few metals occur in pure, or 'native', form. Most are found in **mineral compounds**, or 'ores'. Metals are separated from the host rock by smelting: the rock is heated to the melting point of the metal, which is collected for further processing.
◆ Some metals are used pure. Some are combined to produce **alloys** – metals that are harder than the individual components.

★ 614

All that glisters

Fool's gold, a common name for iron pyrite, is a mineral primarily of iron and sulphur. Its brassy yellow colour has fooled many into believing they have struck gold. Unlike gold, iron pyrite corrodes easily. In the past, it was used to make a spark in wheel-lock guns. Today, it is mostly processed to make sulphuric acid.

FOOL'S GOLD Iron pyrite occurs in the USA and Spain.

Precious metals ❷

Gold, silver and platinum occur in pure form, and have a beauty, rarity and strength that makes them highly valued.

Gold does not tarnish or corrode. It is highly malleable: 10 g (¼ oz) can be beaten into a sheet 3 m (10 ft) wide or spun into a thread 1 km (½ mile) long. It is a good conductor of electricity. Modern uses of gold include jewellery, coinage, international currency standard, religious artefacts, tooth fillings, space visors and electronic circuitry.

Silver, the most common precious metal, shares many of gold's properties, but is lighter and tarnishes easily.

Platinum has a very high melting point and does not tarnish.

FACE OF GOLD A 12th- to 13th-century gold funerary mask from Peru.

Alloys ❸

An alloy is a mixture of two or more metals, and may include non-metals as well.

◆ **Bronze** is an alloy of copper and tin.

◆ **Brass** is an alloy of copper and zinc.

◆ **Steel** is an alloy of iron and carbon. Chromium and nickel are added to make steel rustproof and stain-resistant.

◆ **Super alloys** of nickel, cobalt and other elements are used for spacecraft and jet engines.

STEEL STYLE
Stainless steel kitchenware keeps its sparkle.

TITANIUM CURVES
The Guggenheim Museum in Bilbao has a covering of titanium tiles.

Carats ❺

Precious gems such as diamonds, and the purity of precious metals such as gold, are measured in units called carats. The carat was originally based on the weight of grains or seeds. For measuring the weight of **gems**, 1 carat equals 0.2 g (0.00705 oz). Used in relation to **metals**, 1 carat stands for $1/24$. Pure gold is 24 carats; 18 carat gold is 18 parts gold and 6 parts alloy, to give more strength.

Natural metal-forming processes ❹

Type of deposit	Process	Metals found
Main deposits found in areas of former volcanic and igneous rock-forming activity	Metals separate out and collect together as magma cools.	Uranium, platinum
	Metals dissolved in superhot water and steam percolate through surrounding rock.	Gold, silver, mercury, copper
Placer deposits found in streams	Nuggets released from main deposit by weathering are washed away in streams and deposited where the stream-flow slows. Placer deposits have caused many 'gold rushes', such as the 1848–9 California Gold Rush.	Gold, uranium, tin, platinum. Also diamonds and emeralds
Ocean-floor deposits hydrothermal vents	Minerals in water released from Earth's crust collect on seabed.	Copper, iron, zinc
nodules	Lumps of mineral-rich sediment form on the ocean floor.	Manganese, iron, copper, nickel, cobalt

Hallmarks ❻

The hallmark system, which applies to gold, silver and platinum, represents the oldest form of consumer protection.

A hallmark consists of an Assay Office or Town Mark, a Standard Mark and a Sponsor's or Maker's Mark. It guarantees that the piece has been independently tested and conforms to a legal standard of purity.

ANSWERS

79   **Sunstone** – the sheen is caused by internal reflections

126   **Moonstone** ❷

★ **133**   **B: Diamond** ★

165   **Emerald Isle** – the result of a mild, humid climate

210   **Diamond** – in the title of a short story

211   **Baseball** – the batter stands at one point of the diamond

240   **Australia** ❷

354   *Diamonds Are Forever* – a 1971 James Bond film

364   **Jewels** ❷

365   **Pearl** ❻

507   **Diamonds** ❷

562   **Cullinan** ❺

569   **Emerald** ❸

713   **A: Gemstones** – from Latin *gemma*, for precious stone

812   **A jewel** – with a panther-shaped flaw

844   **Blue** ❷

934   **Green** ❷

983   **Diamonds** ❷

Gems

Core facts ❶

◆ **Gemstones** are minerals valued for their colour, rarity, lustre and crystal structure. Many gems are unusually perfect specimens of common minerals. Minute **traces of impurities** in minerals give gemstones their vivid colours.

◆ Some gems are differently coloured versions of the same parent mineral. **Corundum** that is coloured red is called ruby, but when blue or any other colour it is called sapphire. **Quartz** is the parent mineral to several gems, including amethyst and citrine.

◆ **Diamonds** occur in a range of colours, including yellow, pink and blue.

◆ Some gems also have **industrial uses**.

Gemstones ❷

The **most prized** gems are yellow diamonds, grass-green emeralds, blue sapphires and deep red rubies.

	Greens	Blues	Yellows	Reds/purples	Colourless
Precious gemstones (P) have rarity, beauty, size and durability.	Emerald	Sapphire	Diamond	Ruby	Diamond Opal
Semi-precious stones (SP) have only one or two of the qualities of precious stones.	Garnet Jade Malachite Peridot Tourmaline	Moonstone Turquoise	Cat's-eye Citrine	Amethyst Garnet Jasper Topaz Zircon	

Birthstones ❸

January	Garnet
February	Amethyst
March	Bloodstone or aquamarine
April	Diamond
May	Agate or emerald
June	Pearl or moonstone
July	Ruby or onyx
August	Carnelian or peridot
September	Chrysolite or sapphire
October	Beryl, tourmaline or opal
November	Topaz
December	Turquoise or zircon

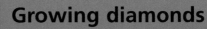

Growing diamonds ❹

Synthetic diamonds can be made in a laboratory by subjecting **carbon** to extreme pressure and heat. Almost any carbon-rich material can be used, including sugar and peanuts. The first synthetic gem-quality diamonds were produced in 1970.

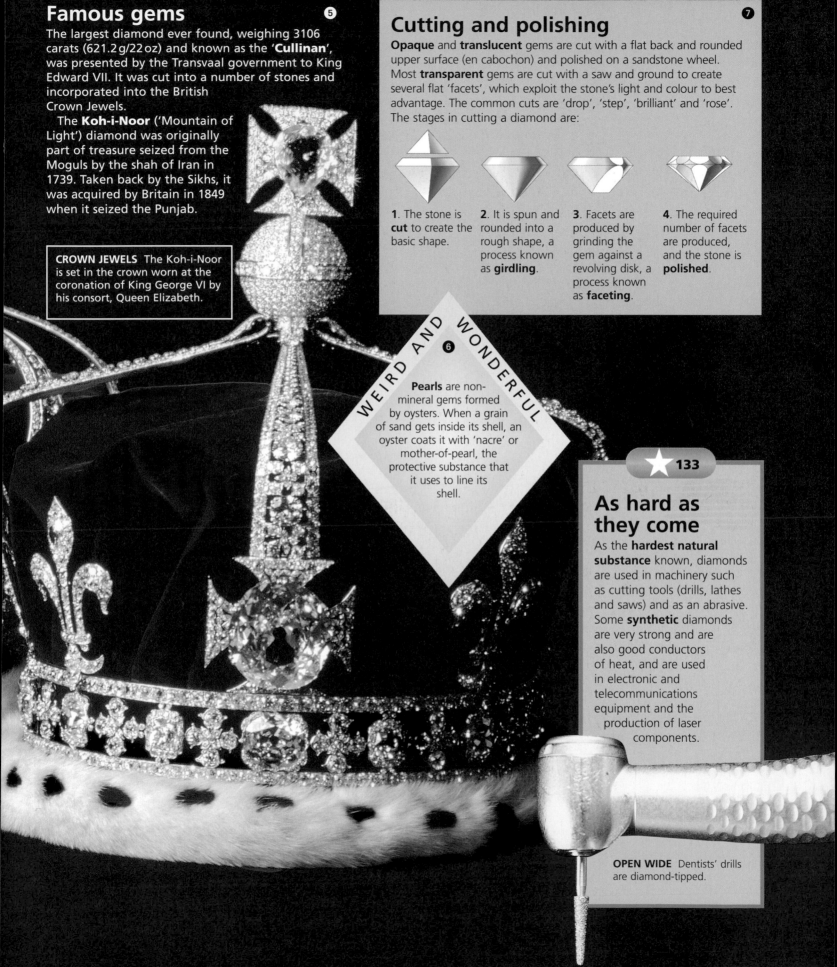

Famous gems

⑤

The largest diamond ever found, weighing 3106 carats (621.2 g/22 oz) and known as the **'Cullinan'**, was presented by the Transvaal government to King Edward VII. It was cut into a number of stones and incorporated into the British Crown Jewels.

The **Koh-i-Noor** ('Mountain of Light') diamond was originally part of treasure seized from the Moguls by the shah of Iran in 1739. Taken back by the Sikhs, it was acquired by Britain in 1849 when it seized the Punjab.

CROWN JEWELS The Koh-i-Noor is set in the crown worn at the coronation of King George VI by his consort, Queen Elizabeth.

Cutting and polishing

⑦

Opaque and **translucent** gems are cut with a flat back and rounded upper surface (en cabochon) and polished on a sandstone wheel. Most **transparent** gems are cut with a saw and ground to create several flat 'facets', which exploit the stone's light and colour to best advantage. The common cuts are 'drop', 'step', 'brilliant' and 'rose'. The stages in cutting a diamond are:

1. The stone is **cut** to create the basic shape.

2. It is spun and rounded into a rough shape, a process known as **girdling**.

3. Facets are produced by grinding the gem against a revolving disk, a process known as **faceting**.

4. The required number of facets are produced, and the stone is **polished**.

WEIRD AND WONDERFUL

⑥

Pearls are non-mineral gems formed by oysters. When a grain of sand gets inside its shell, an oyster coats it with 'nacre' or mother-of-pearl, the protective substance that it uses to line its shell.

⭐ 133

As hard as they come

As the **hardest natural substance** known, diamonds are used in machinery such as cutting tools (drills, lathes and saws) and as an abrasive. Some **synthetic** diamonds are very strong and are also good conductors of heat, and are used in electronic and telecommunications equipment and the production of laser components.

OPEN WIDE Dentists' drills are diamond-tipped.

Energy resources

Core facts ❶

◆ **Fossil fuels** are derived from the organic remains of plants and animals. The principal fossil fuels are coal, oil and gas. They account for about 80 per cent of energy consumption.
◆ Fossil fuels have to be burned to produce power. They are classified as **non-renewable** because existing reserves are used up faster than new ones form.

◆ The refining processes for oil and coal produce useful **by-products**.
◆ **Nuclear** power comes from uranium, a metal that occurs in the Earth's crust.
◆ **Alternative energy** harnesses the power inherent in natural sources, such as the wind, tides and heat from the Sun. Some are renewable and some non-renewable.

Coal ❷

Coal is a type of **sedimentary rock** formed from the remains of plants. It probably began accumulating when plants first appeared 425 million years ago, but most was laid down in the Carboniferous period. Decaying trees in swampy areas formed peat marshes. As the peat became more deeply buried it turned into lignite, then bitumenous coal, and then anthracite – the best-quality coal. Coal provides 27 per cent of the world's energy. It is used in power stations for electricity generation and in industry.

COAL FACE A giant drill cuts into the rock face at a coal mine in Germany.

Oil and gas ❸

Oil and natural gas originate with the build-up of sediment in ocean areas that are rich in marine life. With increasing burial over a period of millions of years, the remains of **microscopic marine creatures** are altered by chemical reaction to produce deposits of oil and gas.

No known sources of oil and gas are less than 1–2 million years old.

Major oilfields exist in the Middle East, especially Saudi Arabia and Iran; in the USA, especially Texas; in Russia; in Central Asia, especially Azerbaijan and Kazakhstan; and in the North Sea.

Nuclear energy ❹

Uranium, a rare element found in the Earth's crust, is used to produce heat through nuclear fission (splitting the uranium atoms). The heat is used to produce steam, which drives turbines to produce electricity. The waste is highly radioactive and must be isolated from all life for thousands of years. Leading users are the USA, western Europe and Japan.

Solar power ❺

◆ **Heat from the Sun** can be used to generate power. The simplest form is a south-facing window.
◆ Roof-mounted **solar panels**, or collectors, are used for heating and hot water. In Israel, solar panels provide hot water for more than 80 per cent of homes.
◆ On a larger scale, arrays of **mirrors** that track the Sun reflect its light onto a tower, where it heats water or steam to power turbines that generate electricity.
◆ Another method uses roof-mounted **photovoltaic cells** that absorb sunlight and convert it directly into electricity. Developing countries such as the Dominican Republic, Sri Lanka and Zimbabwe are leading users of this method.

SUNSEEKERS Rows of mirrors soak up the Sun's energy at the Whitecliffs solar power plant, Australia.

Geothermal energy ❽

The requirements are a natural source of heat such as a magma chamber; underground reservoirs; and porous rock that allows water and steam to circulate. The steam and hot water are used directly for heating. Steam is used for generating electricity.

Leading users are Iceland, where more than 45 per cent of the country's energy comes from geothermal sources, the USA, the Philippines, Indonesia, Mexico, Italy and New Zealand.

HEAT AND STEAM A geothermal power plant in the volcanic area of Rotorua, New Zealand.

Hydroelectric power ❻

This uses the force of **falling water** to power turbines that generate electricity. The water is stored in a reservoir and channelled down through a dam. The requirements for hydroelectric power are a strong, fast flow of water and a significant drop down which the water can fall.

This is a renewable energy source, but dams have a finite lifetime. Leading users include Norway and Brazil, where hydroelectric power accounts for more than 90 per cent of domestic electricity generation, USA, Canada, China and Russia.

ATATURK DAM, TURKEY The largest in a series of 22 dams and 19 hydroelectric stations built on the upper reaches of the Euphrates and Tigris rivers to provide irrigation water and hydroelectricity.

Tidal power ❼

To harness the power of **rising and falling tides** to generate electricity requires a minimum tidal range of 8 m (26 ft) between high and low tides, and a narrow bay or estuary that maximises the force of the tidal flow in and out. The strong in-and-out flow of the tide drives turbines in a dam built across the bay or estuary.

Tidal power is a renewable energy source, but it upsets estuary habitats and few coastlines provide suitable conditions.

The largest tide-powered plant is on the River Rance in France, which produces most of the electricity used in Brittany and some for other regions. There are also experimental facilities in Canada, Russia and China.

★ **222**

Wind power

This uses the power of the wind to turn windmills and operate turbines that generate electricity.

The main requirements for a **wind farm** are a site where the wind blows constantly and from a consistent direction, such as on mountain passes, offshore locations and islands. Leading users include the USA, Germany, Denmark and India.

The continents

Core facts ①

◆ There are **seven** continents based on the major landmasses and their associated islands.
◆ A continent's **boundary** extends to the edge of the continental shelf, which may be several hundred kilometres out to sea.
◆ The **interior** of each continent contains an area of old (Precambrian) rock, known as a shield. These are geologically very stable areas of relatively low elevation.
◆ Russia straddles the boundary between Europe and Asia and does not recognise a dividing line between the two. A political boundary runs down the eastern side of the Ural Mountains and the Kazakhstan border.

Name origins ②

Africa Named by the Romans, after the Arabic word *afar* meaning 'dust' or 'earth'.
America Named for the European explorer Amerigo Vespucci; later latinised to Americus.
Antarctica Means 'opposite to the Arctic'; arktikos is the Great Bear constellation above the North Pole.
Asia Assyrian for 'rising sun', or Sanskrit *usa* meaning 'dawn'.
Australia From *terra australis incognita*, meaning 'unknown southern land'.
Europe From the Phoenician word *ereb* meaning 'evening' or 'west'; or from the princess Europa of Greek mythology.

DEATH VALLEY The lowest point in North America is also one of the world's hottest and driest places.

North America ④

Area 25 349 000 km² (9 784 714 sq miles), or 16.1 per cent of the Earth's total land surface.
◆ Includes Greenland, Central America, and the majority of the Caribbean islands. Hawaii is physically part of Oceania, but politically part of North America.
◆ Climate zones range from tropical rain forest to tundra.
◆ The Great Plains separate the mountain ranges down the east and west coasts.

Mt McKinley, Alaska 6194 m (20 321 ft)

86 m (282 ft) below sea level Death Valley, California

Aconcagua, Argentina 6959 m (22 831 ft)

40 m (130 ft) below sea level Valdez peninsula, Argentina

Continental comparisons ③

Largest landmass	Asia
Smallest landmass	Oceania
Warmest	Africa
Coldest	Antarctica
Longest coastline	Asia
Shortest coastline	Africa
Most populated	Asia
Least populated (excluding Antarctica)	Australia
Most countries	Africa (53)
Most northerly land	Oodaq Island, off North Peary Land, Greenland
Oldest rock yet discovered	Zircon from Western Australia, 4.4 billion years old

South America ⑤

Area: 17 611 000 km² (6 797 846 sq miles) or 12.1 per cent of the Earth's total land surface.
◆ The continent's northern boundary is the Panama-Columbia border, not the Panama Canal.
◆ The most southerly point is Cape Horn on Horn Island. The most southerly point on the mainland is Cape Froward, southern Chile.
◆ The major features are the Andes mountains, the Amazon basin and the Patagonian plateau.

A continental landmass ❻

COASTAL MOUNTAINS Mountain ranges such as the Andes are uplifted when continental and oceanic plates collide.

Ancient Precambrian shield areas occur at the core of continents. The shield is usually overlain by younger, mainly **sedimentary rocks**, although it is sometimes exposed. Mountain ranges at the continent edges are created when oceanic and continental plates collide. **Fold mountains** in the centre of continents mark where two continental plates collided and became welded together.

SURFACE COVERING Younger, mostly sedimentary, rock.

SHIELD Ancient basement rocks at the continent's core.

FOLD MOUNTAINS Ranges such as the Himalayas and the Alps were created when continents collided.

Europe ❼

Area: 10 245 000 km² (3 950 000 sq miles), or 6.8 per cent of the Earth's total land surface.
◆ Mountain ranges in Scandinavia and the Alps in southern Europe are separated by a plain that covers European Russia and spreads west to the Low Countries.
◆ Much of Turkey is in Asia. Its largest city, Istanbul, is split between Europe and Asia.

Asia ❾

Area: 44 493 000 km² (17 174 300 sq miles), or 29.6 per cent of the Earth's total land surface.
◆ Extends from the Arctic Circle to 10° south of the Equator.
◆ The Europe/Asia boundary in the Caucasus is disputed.
◆ The northern forests (taiga) are the greatest region of continuous forest in the world.
◆ Contains the world's highest point, Mount Everest.

⭐ 302

Two Capes

The **Cape of Good Hope** marks the most southerly point of Africa.
    Cape Horn marks the southern extremity of South America. It is within 1000 km (600 miles) of the Antarctic Circle – farther south than New Zealand.

Mt Elbrus, Russia
5642 m (18 510 ft)

28 m (92 ft) below sea level
Caspian Sea

400 m (1312 ft) below sea level
Dead Sea, Israel/Jordan

Mt Everest, China/Nepal
8848 m (29 029 ft)

156 m (511 ft) below sea level
Lake Assal, Djibouti

Puncak Jaya, Irian Jaya
5030 m (16 502 ft)

Mount Kilimanjaro, Tanzania
5895 m (19 340 ft)

Oceania ❿

Area: 8 945 000 km² (3 454 000 sq miles), or 6 per cent of the Earth's total land surface.
◆ Includes Australia and New Zealand (Australasia) and the Pacific island groups of Melanesia, Micronesia and Polynesia.
◆ The boundary with Asia is disputed. Western New Guinea (Irian Jaya) is politically part of Indonesia (Asia), but generally regarded as part of Oceania.
◆ Contains the world's largest living organism – the Great Barrier Reef (2027 km/1256 miles long), and Uluru, the world's largest monolith.

Africa ❽

Area: 30 293 000 km² (11 693 098 sq miles), or 20.2 per cent of the Earth's total land surface.
◆ The boundary between Africa and Asia is considered by most to be the Suez Canal, not the Egypt-Israel border.
◆ Africa stretches from the Mediterranean to the Cape of Good Hope.
◆ Climate zones include grassland, desert and rain forest.
◆ Was joined to Asia until the Suez Canal was built.

16 m (52 ft) below sea level
Lake Eyre, South Australia

Mt Vinson
4897 m (16 066 ft)

2538 m (8325 ft) below sea level
Bentley Subglacial Trench

Antarctica ⓫

Area: 13 975 000 km² (5 394 350 square miles) or 9.3 per cent of the Earth's total land surface.
◆ Consists of two massive blocks of ancient rock separated by a deep channel.
◆ Ninety-eight per cent is covered in ice.
◆ Antarctica was once in the tropics.
◆ Contains Mount Erebus, the world's most southerly active volcano.

MOUNT COOK At 3754 m (12 316 ft) Mount Cook is New Zealand's highest point.

KEY

Highest point on continent

Lowest point on continent

Mountains 1

Core facts

◆ A mountain is defined as an **upward projection** of the Earth's surface with an altitude of at least 600 m (1968 ft).
◆ Most major mountain ranges form where two of the Earth's plates push together, which is also where most volcanoes occur.
◆ The processes that produce mountains are collectively called **orogenesis**, from 'oros'

meaning mountain, and 'genesis', to come into being. There are **three types** of mountain – fold, fault-block, and volcanic (see page 114).
◆ Weathering and **erosion**, especially by glaciers, gradually break up a range and create individual peaks.
◆ Individual mountains that are not part of a range are known as **monarchs**.

Major mountain ranges

Many ranges are part of larger mountain belts. One major belt runs down the west coast of America; another from the Alps and Atlas mountains to the Himalayas and into South-east Asia.

	Main range	Features	Major peaks	Other ranges on continent
	Andes Length: 7242 km (4500 miles)	Longest unbroken range in the world	Aconcagua Cotopaxi Huascaran Llullaillaco	Brazilian East Coast Range Amambai Mountains (Brazil, Paraguay)
	Rockies Length: 4800 km (3000 miles)	Second-longest range in world; longest in North America	Mount Elbert Pikes Peak Windom Mounts Summit Peak	Alaska Range (USA) Appalachians (USA) Cascades Range (Canada, USA) Sierra Nevada (USA)
	Great Dividing Range Length: 3700 km (2300 miles)	Longest range in Australia	Mount Koskiusko	Blue Mountains Australian Alps
	Himalayas Length: 2500 km (1500 miles)	Contain nine of the ten highest peaks in the world, and 30 peaks over 7620 m (25 000 ft)	Everest Kanchenjunga Annapurna	Western and Eastern Ghats (India) Hindu Kush (Pakistan, China, Afghanistan)
	Alps Length: 1207 km (750 miles)	A great arc of fold mountains in Europe	Mont Blanc Matterhorn Eiger Jungfrau	Appenines (Italy) Caucasus Grampians (Scotland) Pyrenees (Spain) Urals (Russia)
	Atlas Length: 2000 km (1200 miles)	Separate Sahara desert and Mediterranean Sea	Jebel Toubkal	Ruwenzori (Uganda, Democratic Republic of Congo) Drakensberg (South Africa)

THE ANDES The range is one of the youngest. It began forming 4.5 million years ago.

★ 151

The Andes

The Andes run almost the entire length of South America. The system is divided into two parallel ranges: the older eastern range, the **Cordillera Oriental**; and the younger, western range, the **Cordillera Occidental**. They are separated by a high-altitude plateau called the **Altiplano**.

Types of mountains ❸

◆ **Fold mountains** form where layers of folded rock are pushed upwards along the edge of a continental plate as two plates collide. Most mountains formed this way, including the Alps, Himalayas and Appalachians.

◆ **Fault-block mountains** are formed by the displacement of rock along a fault in the Earth's crust. The Sierra Nevada in California, the Grand Tetons in Wyoming, and the mountains along the East African Rift Valley formed this way.

Old and new ❺

The Earth's surface is a battleground between the **constant upward force** of mountain building and the **erosional forces** of ice, water and gravity.

As soon as a range forms, weathering begins breaking up the rock surface. Glaciers, rivers and rockfalls all transport this debris away, leaving individual, rocky peaks and ridges, and numerous valleys. Younger ranges such as the Alps and Himalayas, which began forming around 4.5–10 million years ago, are at this stage. As weathering and erosion continue, the peaks are gradually worn down into rounded hills. This can be seen in parts of Scotland and Canada, where mountains are around 460–390 million years old.

The Himalayas and Tibetan plateau ❻

The Himalayan mountains are formed from sedimentary rock that was first laid down in the sea.

Around 40 million years ago, the **Australian-Indian plate** was drifting northwards towards the **Eurasian plate**. Ancient rock on the sea floor between them was gradually pushed upwards to form the beginnings of the Himalayas and the Tibetan plateau. Eventually the Indian plate moved under the Eurasian plate, pushing up the mountains further still.

The plates are still moving towards each other at a rate of a few centimetres each year. And the whole region has risen about 3000m (10000ft) in the last few million years.

CONTINENTAL COLLISION
The Indian subcontinent drifts north and collides with Asia.

MOUNTAINS FORM Layers of compacted sediment on the ocean floor are compressed by this collision and thrust up in folds.

VIEW FROM ABOVE
This satellite image of the Himalayas shows Everest and the plain of Northern India, crossed by tributaries of the River Ganges, which is fed by water draining from the Himalaya range.

Tie-breaker ❹

Q: What point on the Earth's surface is farthest from its centre?
A: Mount Chimborazo in Ecuador. Although only 6267m (20561ft) above sea level, this Andean peak is 2150m (7054ft) farther from the Earth's centre than Mount Everest because of an outward bulge around the Equator.

The numbers or star following the answers refer to information boxes on the right.

ANSWERS

212 **He skied** – down the Lhotse Face, in 1970

318 **Karakoram** (the range in which it is located) ❶

370 **Climbing mountains** – slang for mountaineer

383 **Kilimanjaro** ❶

422 **Mount Ararat** ❷

443 **Mount Fuji** ❶ ❷

446 **Table Mountain** – visible to ships 90 miles (140km) away

523 **B: George Mallory** – died attempting the summit, 1924

571 **Table Mountain** – highest point 1086m (3563ft)

572 ★ **Mount Everest** ★

573 **Mount Rushmore** ❸

575 **Mount Fuji** ❶ ❷

576 **Mount Kilimanjaro** ❶

577 **The Matterhorn** – 4477m (14688ft)

602 **False** (it's the highest in Western Europe) ❶

661 **Mount Rushmore** ❸

663 **The Eiger** – in *The Eiger Sanction*, 1975

664 **Mount Olympus** ❷

668 **Chris O'Donnell** (set on K2 and released 2000) ❶

670 **Mount Sinai** ❷

804 *North by Northwest* (with Cary Grant, 1959) ❸

805 **The North** – first scaled in 1938; more than 50 have died attempting it

Mountains 2

Tallest mountains by region ❶

Peak	Mountain range	Location	Height
Mount Everest	Himalayas	Nepal/China	8848m (29029ft)
K2	Karakoram	Pakistan/China	8611m (28251ft)
Aconcagua	Andes	Argentina	6959m (22831ft)
Mount McKinley	Alaska Range	USA	6194m (20321ft)
Popocatépetl		Mexico	5465m (17928ft)
Mount Elbert	Rockies	USA	4400m (14436ft)
Mount Rainier	Cascade Range	USA	4392m (14409ft)
Mount Kilimanjaro		Tanzania	5895m (19340ft)
Mount Elbrus	Caucasus	Russia	5642m (18510ft)
Mont Blanc	Alps	France/Italy	4807m (15771ft)
Puncak Jaya (Nnga Pulu)		Indonesia	5030m (16503ft)
Mount Fuji		Japan	3776m (12388ft)
Mount Cook		New Zealand	3754m (12316ft)
Mount Vinson	Vinson Massif	Antarctica	4897m (16066ft)

EVEREST ATTEMPT George Mallory and Edward Norton on Everest's north-east ridge on May 21, 1922.

★ **572**

High points

Mount Everest's Tibetan name *Chomolungma* means 'Mother Goddess of the World'. It was given the name Everest in 1865 after former Surveyor General of India, Sir George Everest (pronounced 'Eve-Rest'). In 1923 British mountaineer George Mallory said he would climb Everest 'because it's there'. But it was not until May 29, 1953, that New Zealander Edmund Hillary and Tenzing Norgay of Nepal first reached its summit.

The world's second-highest mountain is Mount Godwin-Austen in the Karakoram range of the Himalayas – better known as '**K2**'. The **Eiger** in the Swiss Bernese Oberland is known as 'the meanest mountain on Earth', because of the challenge that its north face offers even the most experienced climbers.

Sacred mountains ❷

◆ **Mount Ararat**, the highest peak in Turkey, is traditionally associated with the mountain upon which Noah and his ark landed. According to Persian legend, Ararat is the cradle of the human race.

◆ **Mount Sinai** is associated with the mountain where God revealed the Ten Commandments to Moses. It is sacred in Christian, Jewish and Islamic tradition.

◆ **Mount Olympu**s is the highest mountain in Greece. It lies on the border between Macedonia and Thessaly. In ancient Greek mythology it was the home of the gods, where Zeus viewed the affairs of humans. It is comparable to Asgard, the mythical home of the Norse gods.

◆ **Mount Fuji** is the highest mountain in Japan. Also called Fujiyama, its name means 'everlasting life'. Mount Fuji is sacred in Japan: to climb it is a religious act. The Japanese artist Hokusai immortalised it in his print series *Thirty-six views of Mount Fuji* (1826-33).

SYMBOLIC MOUNTAIN An artist's interpretation of the ark perched on Mount Ararat (left). Expeditions have failed to find any evidence there.

MYTHICAL HEAVEN In the *Iliad*, Zeus addresses the lesser gods from 'many-ridged Olympus' (below).

FAMILIAR LANDMARK Mount Fuji (below) is a dormant volcano that last erupted in 1707. It is regarded as a sacred place.

THE MATTERHORN English mountaineer Edward Whymper first climbed the Matterhorn, in the Alps, in 1865.

Stone faces ❸

Mount Rushmore in South Dakota is best-known for the 60 m (200 ft) sculptures of (left to right) George Washington, Thomas Jefferson, Theodore Roosevelt and Abraham Lincoln carved in the granite of its north-east face. The memorial was suggested by historian Doane Robinson and carved by John Gutson Borglum.
It was completed in 1941.
Cary Grant and Eva Marie Saint were chased across the memorial in the film *North by Northwest* (1959).

MONUMENTAL FACES The figures can be seen from 97 km (60 miles) away.

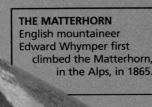

Lakes

Core facts ❶

◆ A lake is a body of water that is **completely surrounded** by land. A lake forms when water fills a depression in the land. Many are relatively short-lived.
◆ Lakes can be **formed by** glaciers, shifting continents, collapsing volcanoes, or rivers with no outlet to the sea.
◆ Lakes can be fresh or salt water.
◆ The **colour** of a lake can be influenced by minerals or algae in the water.

Lake formation ❷

Cause	Examples
Glaciers deepen valleys	Great Lakes, Italian Lakes, Lake Geneva, fjords, lochs
Ice carves out a circular hollow high in mountains	Known as cwms in Wales, cirques in France, tarns in England, corries in Scotland
Glacier carves out depressions over a large area	Known as shield lakes, these occur in Canada and Finland
Moving plates in the Earth's crust create a rift	Dead Sea, Lake Malawi, Lake Baikal
Two continents collide	Caspian Sea
Volcano cone collapses	Crater Lake
River drains into a basin	Lake Chad, Lake Eyre, Great Salt Lake

Salty lakes ❸

The **Dead Sea** is a true salt lake as it formed in a depression into which large quantities of salt and mineral-rich sediment were deposited by streams.
The **Caspian** and **Aral** Seas were once linked to the Mediterranean. Whereas the Caspian is fed by several rivers, and now has brackish rather than true salt water, rivers feeding the Aral Sea have been diverted for irrigation and the sea is shrinking.

SHRINKING SEA Evaporation of the Aral Sea has exposed large areas of salt-ridden seabed, which is being eroded by the wind.

VOLCANIC LAKE Acids have turned the water vivid green in this volcanic crater lake in Costa Rica.

WEIRD AND WONDERFUL ❹
Trinidad's Pitch Lake is filled with a tar-like kind of crude oil that appears on the surface. The lake is around 60 m (200 ft) deep. In some places the pitch is quite liquid; in others, evaporation has hardened the surface.

Lake records ❺

Largest
The Caspian Sea, with an area of 370 998 km² (143 243 sq miles). It is the remnant of an ocean, and is fed by the Volga River.

Deepest
Lake Baikal, around 1640 m (5380 ft) deep. It is the largest freshwater lake by volume.

Highest navigable
Lake Titicaca, on the Peru/Bolivia border. It is in the Andes, 3810 m (12 500 ft) above sea level. It is also South America's largest lake.

Groups of lakes ❻

◆ The **Great Lakes** of the USA and Canada – Superior, Michigan, Huron, Erie and Ontario – are all that remain of a giant ice sheet that covered North America during the last Ice Age. The group forms the largest freshwater surface in the world.
◆ The **Italian Lakes** – Garda, Maggiore, Como, Iseo and Lugano – were cut into the Alpine foothills by glaciers.
◆ The **Lake District** of Cumbria, England, famous for its scenery and its 19th-century poets, was carved out by ancient glacial action. Its lakes include Windermere (England's largest), Ullswater, Coniston Water, Derwentwater and Wastwater.

PRE-ICE AGE River valleys in the Great Lakes area.

ICE AGE A sheet of ice covers the region.

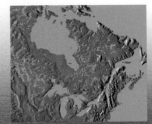
POST ICE AGE Retreating ice leaves large depressions that fill with water.

⭐ 931

Coloured lakes

Minerals and **algae** (microscopic plants) in water can affect a lake's colour. Examples of algae turning water a different colour include Lake Natron in East Africa and Laguna Colorado ('coloured lagoon') in the Bolivian Andes, which are pink due to the presence of algae.

In volcanic regions minerals leech into the lakes and colour the water shades of yellows, blues and greens.

PINK PUZZLE The intense pink colour of Lake Hillier in Western Australia (below) is not due to algae, but the real cause is not known.

Elusive monster ❼

Loch Ness is the UK's longest and deepest lake – nearly 39 km (25 miles) long and up to 240 m (800 ft) deep. Like some deep waters in Scandinavia, Loch Ness is believed by some to contain an aquatic creature, perhaps a modern relative of the prehistoric plesiosaur. The loch is so deep that it is impossible to explore effectively, even using a submersible or satellite. Despite reported sightings and blurred photography, there is no proof of 'Nessie's' existence.

Deserts

Core facts ❶

◆ The **word desert** comes from the Latin for 'abandoned', or 'unoccupied', and refers to large dry regions with sparse vegetation.
◆ Most deserts have **less than 30 cm** (12 in) of rain a year, and it all falls at once.
◆ The main **desert belts** are in the Tropics, where the air is hot and dry. Others form in the interiors of continents far from the influence of the sea. Rainshadow deserts form in the lee of mountain ranges, where rain falls on the windward side.
◆ The majority of **landforms** in deserts were formed by water. The wind moves and deposits large quantities of loose sand to create dunes.

Deserts around the world ❷

Africa
1 Sahara	9 000 000 km² (3 500 000 sq miles)	Largest in the world
2 Kalahari	520 000 km² (201 000 sq miles)	
3 Namib	310 000 km² (120 000 sq miles)	Oldest desert in the world
4 Nubian	260 000 km² (100 000 sq miles)	

Asia
1 Arabian	1 300 000 km² (502 000 sq miles)	Largest in Asia
2 Takla Makan	1 245 000 km² (327 000 sq miles)	
3 Gobi	1 040 000 km² (401 000 sq miles)	
4 Turkestan	450 000 km² (175 000 sq miles)	

North America
1 Great Basin	492 000 km² (190 000 sq miles)	Largest in North America
2 North Mexico	450 000 km² (175 000 sq miles)	
3 Sonoran	260 000 km² (100 000 sq miles)	
4 Mohave	65 000 km² (25 000 sq miles)	

South America
1 Patagonian	673 000 km² (260 000 sq miles)	Largest in South America
2 Atacama	140 000 km² (54 000 sq miles)	

Australia
1 Great Victoria	647 000 km² (250 000 sq miles)	Largest in Australia
2 Great Sandy	400 000 km² (150 000 sq miles)	
3 Simpson	145 000 km² (56 000 sq miles)	

★ 180

The desert floor

Not all deserts are sandy. In some, such as the **Dasht-e-Lut** in eastern Iran, the desert surface has been worn down to bare rock, known as hamada.

In others, such as the **Sonoran**, the wind has removed all the finer sediment, leaving a floor of compacted pebbles or gravel, called desert pavement.

Tie-breaker ❸

Q: The British explorer St John Philby was the first European to cross which desert?
A: The Empty Quarter (Rub' al-Khali) of Arabia.
Philby crossed the desert from The Gulf coast to the Red Sea in 1917, as part of a diplomatic mission to a local ruler.

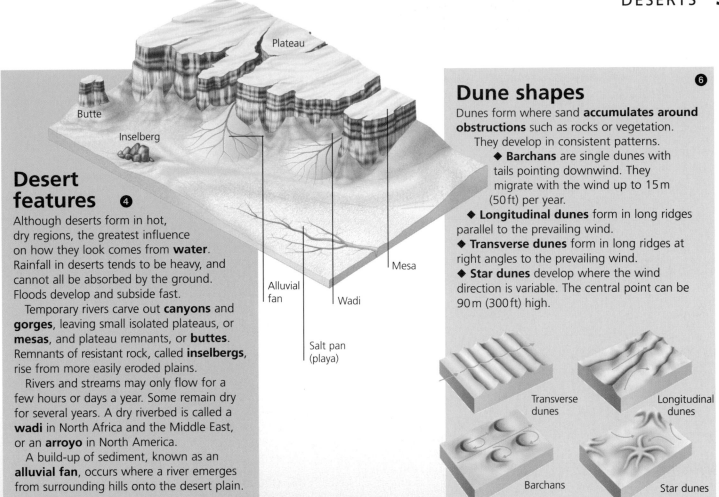

Plateau

Butte

Inselberg

Mesa

Alluvial fan

Wadi

Salt pan (playa)

Desert features ❹

Although deserts form in hot, dry regions, the greatest influence on how they look comes from **water**. Rainfall in deserts tends to be heavy, and cannot all be absorbed by the ground. Floods develop and subside fast.

Temporary rivers carve out **canyons** and **gorges**, leaving small isolated plateaus, or **mesas**, and plateau remnants, or **buttes**. Remnants of resistant rock, called **inselbergs**, rise from more easily eroded plains.

Rivers and streams may only flow for a few hours or days a year. Some remain dry for several years. A dry riverbed is called a **wadi** in North Africa and the Middle East, or an **arroyo** in North America.

A build-up of sediment, known as an **alluvial fan**, occurs where a river emerges from surrounding hills onto the desert plain.

Dune shapes ❻

Dunes form where sand **accumulates around obstructions** such as rocks or vegetation. They develop in consistent patterns.

◆ **Barchans** are single dunes with tails pointing downwind. They migrate with the wind up to 15 m (50 ft) per year.
◆ **Longitudinal dunes** form in long ridges parallel to the prevailing wind.
◆ **Transverse dunes** form in long ridges at right angles to the prevailing wind.
◆ **Star dunes** develop where the wind direction is variable. The central point can be 90 m (300 ft) high.

Transverse dunes

Longitudinal dunes

Barchans

Star dunes

Desert rain ❺

On the rare occasions when rain falls, deserts can be transformed overnight into a sea of colour. Seeds that have been dormant since the previous rains spring into life, carpeting the desert floor with grasses and flowering plants. The plants have to complete their lifecycle and produce new seeds before the moisture dries up. Other plants have waxy surfaces that reduce moisture loss; they sit out the heat until rains arrive, then they burst into flower.

BEFORE AND AFTER
The Namib Desert is transformed by the arrival of the rains.

Glaciers and ice sheets

Core facts ❶

◆ A glacier is a thick mass of **accumulated ice** that moves, or flows, downhill under the force of gravity. There are two main types: ice sheets and valley glaciers.

◆ In an **ice sheet**, ice flows out in all directions from one or more snow-accumulation centres and covers all the underlying land. Greenland and Antarctica currently qualify as ice sheets.

◆ Where the tongue of a glacier extends out over the sea it forms an **ice shelf**.

◆ **Valley glaciers** occur in mountains, where they form streams of ice flowing downhill between rock walls. As ice flows, it erodes and alters the landscape beneath it.

◆ The Arctic Ocean is covered by large areas of floating or **pack ice**.

Major glaciers in each region ❷

Antarctica Lambert-Fisher 515 km (320 miles); Arctic Institute 418 km (260 miles); Beardmore 225 km (140 miles)

Russia Novaya Zemlya 418 km (260 miles)

USA Bering (Alaska) 204 km (126 miles); Hubbard (Alaska) 146 km (91 miles)

Greenland Petermann 200 km (124 miles); Humboldt 114 km (71 miles)

Norway Jodestals 75 km (45 miles)

India/Pakistan Siachen 70 km (44 miles)

Antarctic ice sheet ❸

◆ Antarctica is the **largest ice sheet** in the world, covering an area of 13.9 million km² (5.4 million sq miles), one and a half times the area of the USA.

◆ The ice sheet is up to 4300 m (14,000 ft) thick in places and contains 80 per cent of the world's ice. Two-thirds of the world's freshwater is locked up in it.

◆ Where no ice has melted at the surface of the underlying bedrock, the ice can be 1 million years old.

◆ Antarctic glaciers are the largest in the world and include the Lambert-Fisher glacier, which drains one-fifth of the Antarctic ice sheet.

◆ Twelve per cent of all the freshwater on Earth passes through the Lambert-Fisher glacier.

RIVER OF ICE The Hubbard – one of the fastest-moving glaciers in North America – empties into Yakutat Bay, Alaska.

Ice shelves ❹

Along parts of the Antarctic coast, glacial ice flows into bays, where it forms shelves of floating ice hundreds of metres thick. They extend out to sea, but remain attached to land on at least one side. Shelves can remain stable for millions of years. The **Ross Ice Shelf** is the largest, and covers a similar area to France. The **Filchner** Ice Shelf is the second largest.

Greenland

❺

◆ The whole interior of Greenland is covered by a vast **ice sheet** that is a remnant of the last Ice Age. It covers an area of 1.7 million km^2 (64872 million sq miles), or 82 per cent of the island.

◆ The ice is an average 1500 m (5000 ft) thick, and reaches a maximum 3000 m (10000 ft) in places.

◆ Greenland is the main source of **icebergs** in the Northern Hemisphere. It releases more than 5000 icebergs a year into Baffin Bay, including the one hit by the *Titanic*.

Icebergs

★ 534

Icebergs are fragments of ice sheets and glaciers that formed on land. Where a glacier reaches the sea, pieces break off the leading edge, or 'calve'. They float low in the water with about 80 per cent submerged.

Arctic icebergs average about 180 m (660 ft) long and 45 m (150 ft) high. The largest recorded iceberg in the Northern Hemisphere was 11 km (7 miles) long and 5.9 km (3^1/$_2$ miles) wide, seen in 1882.

In **Antarctica**, huge fragments break off the edges of ice sheets to form **tabular** icebergs, sometimes over 95 km (60 miles) long. The largest known was 8000 km^2 (3080 sq miles). They break up once at sea.

The tallest recorded iceberg was 134–158 m (447–527 ft) high, measured from a helicopter in the 1950s.

EXPOSED
Seawater has eaten away at the area that was underwater.

Valley glaciers

❻

Valley glaciers come in all shapes and sizes. They follow valleys originally occupied by streams, and some have branching tributaries. The **weight** of the ice is so great that it scours, scrapes and tears out rock from the valley floor and walls, and carries this load, called **moraine**, down the mountain. This debris is deposited when the glacier retreats again.

Main glaciers erode more heavily than tributaries because of the great weight of ice. The main valley floor is eroded into a typical **U-shape** and to a greater depth than tributary valleys. These are left as **hanging valleys** when the glacier has receded. Sharp angular crests and divides between valleys are called **aretes**. Small lakes called **corries** form in scoured-out depressions.

The leading edge of a glacier is called the **snout**.

When a glacier recedes, meltwater sometimes forms a lake, and long, low hills called **drumlins** are revealed.

Arete

Corrie

Hanging valley

U-shaped valley

Snout

Lake

Drumlins

Pack ice

❼

Pack ice forms from **seawater**. It is broken into slabs by channels of broken water and moves around. Melting occurs during spring and summer.

The **Arctic ice field** consists of 4.7 million km^2 (1.8 million sq miles) of 3–6 m (10–20 ft) thick ice that never melts. The pack ice and seasonal sea ice extend into the nearby North Atlantic.

Twice as much pack ice forms around Antarctica as in the Arctic, but only a few areas have permanent ice.

Rivers 1

Core facts ❶

◆ **River systems** collect surface water and channel it back to the sea, to a lake or, in a few cases, into a desert (see pages 90–1).
◆ A river system includes the main channel and all the tributaries that flow into it.
◆ Water in rivers comes from rain, or meltwater from snow and ice. On impermeable ground rainwater runs off and collects in **streams**. On permeable ground, it sinks in and emerges as **springs** at a lower level.
◆ The area surrounding a river and its tributaries is called a **drainage basin**. Drainage basins can be huge, and are separated from each other by high ground called **watersheds**.

From source to mouth ❷

Drawn downwards by **gravity**, water finds the **shortest route** downhill. Where it meets weaknesses in the underlying rock it cuts down to form waterfalls, gorges and valleys. When a river reaches flatter ground, its slows down. Some flow in wide, curving meanders as they cross broad flood plains.

1 SOURCE The place where the sources come together to form a single flow is called the **headwater**.

2 UPPER REACHES The current is strong and a river cuts channels through exposed soft rock. **Waterfalls** form where harder rock meets soft.

CASCADES A stream finds the fastest route downhill.

3 LOWER REACHES On lower, flatter ground the river flow slows. On flat flood plains it forms wide loops, or **meanders**.

FLOOD PLAIN Loops and oxbow lakes have formed where meanders became cut off from the main channel.

4 MOUTH A river **discharges** into the sea or a lake.

Perennial and seasonal ❸

◆ **Perennial** rivers flow all year. They are fed by an **unlimited water supply**, and keep flowing even in desert areas if the catchment area lies in a region with a wetter climate. Examples are the Nile and the Colorado.
◆ **Seasonal** rivers are those that tend to dry up in summer. They usually occur in areas that have a wet winter and dry summer. Examples include the River Fincke in Australia and some rivers of southern Europe.

238

Niagara Falls

The Niagara Falls lie midway along the Niagara River, which flows between lakes Ontario and Erie on the border between USA and Canada. The name 'Niagara' comes from the Iroquois for 'neck of land cut in two'. The Falls are divided by Goat Island into the Horseshoe Falls (670 m/ 2200 ft wide) on the Canadian side and the American Falls (320 m/ 1050 ft wide).

The Falls date back about 10 000 years to the end of the last Ice Age. Since then, they have moved 11 km (6³/4 miles) upriver as the Niagara cliffs are eroded. The Falls should reach Lake Erie in 25 000 years' time.

GIANT FALLS
The Horseshoe Falls carry about 90 per cent of the 5.5 million litres (1.2 million gallons) of water that flows over Niagara Falls each second.

Waterfalls

There are **three types** of waterfall. A **cataract** is a high fall over which large volumes of water pass. A **cascade** is a low (and usually less steep) fall. **Rapids** are a turbulent flow of white water caused by a steep-sided channel.

	Location	Features
Largest volume of water	Stanley Falls, Lualaba river, Congo	17 000 m³ (600 900 cu ft) per second. Named after the explorer Sir Henry Stanley.
Widest in the world	Khone Falls, Mekong river, Laos	10.6 km (6³/4 miles) wide. Second-largest by volume of water; nearly twice that of Niagara Falls.
Highest in the world	Angel Falls, Churun river, Venezuela	Highest single drop is 807 m (2650 ft). Named for Jimmy Angel, who crashed his plane nearby in 1937.
Highest in Europe	Utigård, Nesdale, Norway	800 m (2625 ft)
Most famous in literature	Reichenbach Falls, Switzerland	Where Sherlock Holmes and Moriarty fell to their deaths.
Most famous discovery	Victoria Falls, Zambezi river, Zambia and Zimbabwe	Local name Mosi-oa-Tunya: 'Smoke that Thunders'. Named by Dr Livingstone for Queen Victoria.

ANGEL FALLS The Churun river plunges over the edge of the plateau.

Daredevil stunts

In 1859 French acrobat **Charles Blondin** (real name Jean-François Gravelet) crossed the American Falls at Niagara on a **tightrope** 335 m (1100 ft) long. He subsequently crossed with his agent on his back, blindfolded, on stilts, and once cooked an omelette halfway across. In 1901 **Annie Taylor** was the first person to go over the Falls in a **barrel**. She survived, as did Bob Leach from Cornwall, England, in 1911. Peter Debernadi and Geoffrey Petkovich were the first to go over in the same barrel and survive.

River of song

The **Mississippi** has been immortalised in song: it is the 'Ol' Man River' in Jerome Kern's musical *Show Boat* (1927), and also Henry Mancini's 'Moon River', which became the theme song to the film *Breakfast at Tiffany's* (1961) with Audrey Hepburn.

Rivers 2

River records ❶

**Africa:
Nile**
Length: 6700 km (4160 miles), the longest river in the world. It flows south to north into the Mediterranean. It drains one-tenth of the African continent and floods seasonally.

The **Zaire** (formerly the Congo) is Africa's second longest river at 4700 km (2900 miles), followed by the **Niger** at 4100 km (2550 miles).

**South America:
Amazon**
Length: 6400 km (3980 miles), the second-longest in the world and largest by volume of water. Rises in the Andes in Peru. Freshwater from the Amazon stretches 180 km (110 miles) into the South Atlantic.

The **Madeira**, a major tributary of the Amazon, is the world's longest tributary at 3380 km (2100 miles).

**Northern Asia:
Chang Jiang (Yangtze)**
Length: 6300 km (3910 miles), the world's deepest river, third longest and the longest in Asia. It rises in Tibet and flows east into the Yellow Sea.

The **Yenisei** is the second longest in northern Asia at 5540 km (3440 miles), followed by the **Huang He** (Yellow River) at 5464 km (3400 miles).

**Southern Asia:
Mekong**
Length: 4180 km (2600 miles), the longest Asian river south of the Himalayas. Rises in central Tibet and flows south-east into the South China Sea.

The **Brahmaputra** is the second-longest (2900 km/1800 miles); followed by the **Indus** (2880 km/1790 miles), and **Ganges** (2500 km/1550 miles).

**North America:
Mississippi-Missouri-Red Rock**
Length: 6000 km (3700 miles), the longest river in North America and fourth longest in the world. Drains 31 US states and two Canadian provinces before entering the Gulf of Mexico, in a shape Mark Twain compared to a 'long, pliant apple-paring'.

**Australia:
Murray-Darling**
Length: 3370 km (2190 miles), the longest river in Australia. It drains an area of more than 1 million km² (386 000 sq miles). Flows inland from the Great Dividing Range via Lake Alexandrina to Encounter Bay in the Indian Ocean. The Murray and the Darling join at Wentworth in Victoria.

**Eastern Europe:
Volga**
Length: 3700 km (2300 miles), the longest river in Europe. It flows into the Caspian Sea and its basin spreads across about 40 per cent of European Russia.

The **Danube** is the second longest (2820 km/1755 miles) and the only major east-flowing river in Europe. It passes through eight countries and five capital cities before flowing into the Black Sea.

**Western Europe:
Rhine**
Length: 1300 km (800 miles), the longest river in Western Europe. Rises in the Swiss Alps and flows through six countries to empty into the North Sea.

The **Shannon** in Ireland is the longest river in the British Isles (386 km/240 miles).

The **Thames** rises in the Cotswolds and flows 338 km (210 miles) via London into the North Sea.

Crossing the Rubicon ❷

In 49 BC **Julius Caesar** led his army from Gaul across the Rubicon river and took Rome. By so doing he broke the law that forbade a Roman general to lead an army out of his own province, and committed himself to a civil war against the Senate and rival general **Pompey**. 'To cross the Rubicon' came to mean 'to commit oneself to an irrevocable action'.

Name origins ❸

Mississippi, Thames, Nile, Rhône, Niger	Like many of the world's rivers, the original dialect name of these simply means 'river' or 'big river'
Amazon	Named after legendary female warriors thought to have lived on its banks
Congo	Bantu word meaning 'mountain', after local terrain
Huang He	Chinese for 'yellow river' – its European name – referring to its yellow silt
Limpopo	Corruption of local name Lebempe, meaning 'dark river' or 'crocodile river'
Murray-Darling	First named the Hume after one of its explorers, renamed after Colonial Secretary Sir George Murray and New South Wales governor Sir Ralph Darling
Orange	Named in honour of the Dutch royal House of Orange (which included William III of England)
Potomac	From the Indian word *Patawomeck* meaning 'where goods are brought in', referring to a trading post there
Suwannee	Originally Guasaca Esqui (river of reeds), Suwannee is thought to be an African-American corruption of San Juanee (little St John)
Tiber	After Tiberinus, king of a region south of Rome called Alba Longa, who drowned in the river

WEIRD AND WONDERFUL ❺

Every second, the **River Amazon** discharges an average of 200 000 m³ (7 063 000 cu ft) of water into the South Atlantic. This is enough water to fill more than 200 Olympic-sized swimming pools.

⭐ 18

The Nile

The Nile begins as two rivers, the White Nile and the Blue Nile. The **White Nile**'s farthest headstream is the Kagera River in Burundi, which flows into Lake Victoria. Water leaves the lake at the Rippon Falls, which are the true start of the White Nile. It flows north to Khartoum in the Sudan, where it is joined by the **Blue Nile**, which rises in the Ethiopian Mountains.

Hudson River ❹

The Hudson rises in the **Adirondack Mountains**, New York State, and flows 485 km (300 miles) south to New York City. In the 19th century it was linked by canals to the Great Lakes and the St Lawrence and Delaware rivers, and was central to the development of New York City and the American Midwest. It is still one of the most important waterways in the world.

SHIPPING ROUTE The lower Hudson is navigable to ocean-going shipping.

SOMETHING TO SING ABOUT
The Suwannee River (background) rises in the Okenofee swamp in the state of Georgia (USA), then meanders across northern Florida on its way to the Gulf of Mexico. It is the famous Swanee River of the song 'Old Folks at Home'.

Deltas and estuaries

Core facts   ❶

◆ A **delta forms** where a river's flow is checked as it meets the sea and the river deposits its sediment load at the river mouth.

◆ There are two principal types of delta: a **fan** delta and a **bird's foot** delta.

◆ If deposited sediment creates a barrier, a river will break its banks farther upstream and shift its course to the sea.

◆ **Estuaries** form where there are depressions around or barriers across river mouths.

Delta types   ❷

In the case of a **fan delta**, such as the Nile, the river has divided into several channels that fan out, depositing sediment across a large area. Strong wave action redistributes sediment along the coast.

A **bird's-foot delta**, such as the Mississippi, builds up where sediment is deposited in a small area at the delta front, creating an extension. The Mississippi has shifted its course many times, creating a series of subdeltas.

NILE DELTA The Nile begins splitting into several channels more than 160 km (100 miles) inland, and fans out across the entire delta area.

MISSISSIPPI DELTA The river is confined to its main channel for most of its length. Sediment is deposited at the delta front and extends into the sea.

Delta facts and records   ❸

River and location	Features
Brahmaputra/Ganges Bay of Bengal	The world's **largest** delta (60 000 km²/23 160 sq miles). Since 1785, the Brahmaputra river has moved 65 km (40 miles) to the west.
Huang He (Yellow River) Yellow Sea	The world's **muddiest** river flows into the world's **fastest-growing** delta. The river mouth has moved several times due to sediment build-up.
Neva Baltic Sea	**St Petersburg** is spread across more than 40 islands in the delta, hence its nickname 'Venice of the North'. Like Venice, it floods badly.
Orinoco Atlantic Ocean	The delta is 440 km (270 miles) wide. It is rich in alluvium, but has poor drainage. Columbus reached it on his fourth voyage (1498–1500).
Rhône Mediterranean Sea	A **wave-dominated** delta 40 km (25 miles) long. The Grand and Petit Rhône enclose the Camargue region. Contains large alluvium deposits.
Tiber Mediterranean Sea	The ancient city of Ostia at the mouth of the Tiber was the site of the famous **Roman lighthouse** (AD 50).
Volga Caspian Sea	This fertile, **low-tidal delta** is the largest in Russia (11 730 km²/4530 sq miles). It is a major nature reserve.
Chang Jiang East China Sea	Alluvium deposits (430–500 million tonnes per annum) **extend** the delta 1.6 km (1 mile) into the sea every 100 years.
Yodo Inland Sea	A fan-shaped delta empties into a shallow, sheltered bay. **Osaka** is built on reclaimed delta land.

642

The Camargue

This sparsely populated region of the Rhône delta in south-east France consists of 780 km² (300 sq miles) of **marshland** and **lagoons**. The region is famous for roaming herds of black bulls and wild horses. Salt is exploited in pans between the Vaccarès lagoon and the river. Much of the area is a nature reserve.

Inland delta ❹

The **Okavango River** rises in Angola and flows south-east. But on its journey towards the sea it disappears into the Kalahari desert, where it forms one of the largest delta systems in the world and the largest inland delta, covering 15 993 km² (6175 sq miles).

Each year the river is in flood and spreads through a maze of channels and reed beds, where 95 per cent of the water evaporates.

In years of heavy rainfall, the delta can expand to cover 22 000 km² (8500 sq miles), turning the desert into a shimmering expanse of water where aquatic creatures rub shoulders with desert-dwellers.

DELTA WILDERNESS The Inland delta of the Okavango River spreads across the Kalahari in a wilderness of swamps, islands and reedbeds.

Estuaries

Estuaries form in several ways:
◆ Some, such as the Thames and Severn in England, form when **sea levels rise** and drown an existing river valley. This type is extensive and shallow.
◆ **Wave action** can create a partial barrier of sand across a river mouth, such as the Waddensee in Holland.
◆ Movement in the Earth's crust, caused by an earthquake or volcano, can create a **depression** in an area of coastline. Land surrounding a river sinks and floods, becoming a bay with a narrow opening to the sea. San Francisco Bay and the St Lawrence Estuary formed this way.

MORECAMBE BAY, ENGLAND Five rivers feed into this estuary, famous for its birdlife.

Wetlands

Deltas, estuaries and river flood plains support a wide diversity of wetlands – transitional zones between ocean or river environments and dry land.

Marshes are areas of poorly drained land. In tidal marshes, conditions change with every tide. Marshes also occur inland where groundwater, rivers or lakes cause frequent flooding.

Swamps differ from marshes in that they have waterlogged soils or are flooded for most of the growing season. They are often dominated by tree species, including mangroves in tropical regions. The largest mangrove swamp in the world is found in the Ganges delta.

Flood-plain forests have developed around many rivers including the lower Mekong in Cambodia, and the lower reaches of the Mississippi.

Wetlands are vital in **flood control**, as they act like sponges, absorbing and storing rainwater, and then releasing it slowly, thus reducing peak flood levels. Wetland vegetation helps to stabilise riverbanks and shorelines.

Coastlines

Core facts ❶

◆ The world's coastlines are **shaped by** the forces of inland erosion, mountain-building and glaciation. They are constantly being modified by the action of waves and currents, which erode material in some places and deposit it in others.

◆ **Waves** provide most of the energy that erodes and alters coastlines. They do most damage during storms, when each breaking wave can throw thousands of tonnes of water against the shore.

◆ **Sand** makes up around 75 per cent of the world's ice-free coastlines. Tidal action and currents combine to move millions of tonnes of sand along shorelines, a process known as longshore drift.

Coastal features ❷

The sea **erodes** a coast in stages. Waves wear away softer rock, creating cracks and openings. These are gradually hollowed out and enlarged to create **caves**. Cave roofs may be pierced, forming blowholes.

Where caves have formed back to back on a headland, they may eventually unite to form a **sea arch**. Eventually the roof of the arch collapses, leaving a **sea stack** offshore.

Waves carry sediment along a shore until the current meets deeper water, and the sediment is **deposited**. This action can extend a beach across an adjacent bay to form a **spit**, or form a ridge of sand, known as **tombolos**, linking the mainland to an island.

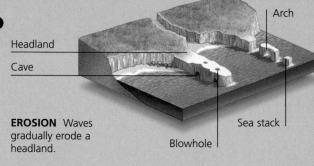

Headland
Cave
Arch
Sea stack
Blowhole

EROSION Waves gradually erode a headland.

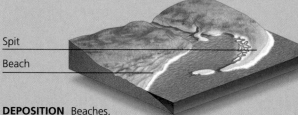

Spit
Beach

DEPOSITION Beaches, dunes and spits are created where sediment is deposited.

THE TWELVE APOSTLES ❸
These giant offshore sea stacks have been carved from the crumbling limestone cliffs of Australia's southern coast. Some of the original twelve have collapsed completely.

Golden sands ❹

More than 90 per cent of beach sand comes from **inland erosion**, deposited by rivers and distributed along coastlines by the action of waves and currents. Most sand on mainland beaches consists of quartz (silica), which is responsible for its beige colour. Some comes from the skeletons and shells of marine creatures, or from the erosion of cliffs. On **volcanic islands**, sand is often black. In **tropical** areas corals and shells produce pink or white sand.

CONCRETE COAST A sea wall on the Gulf of Mexico holds off the encroaching power of the waves.

Coastal defences ⑤

Several methods are used to try to control the natural migration of sand along a shoreline and to mitigate the destructive action of waves. **Jetties** are built at entrances to rivers and harbours to interrupt the movement of sand and to prevent shipping channels filling up.

Groins are built at right angles to the shoreline to trap sand and stop it moving along a beach. **Breakwaters** are built parallel to the shore to create quiet areas near the shoreline. **Sea walls** are large structures built along a shoreline to protect the areas behind from the destructive power of large waves.

★ 341

Lagoons

Lagoons form along coasts where **low ridges of sand** – known as barrier islands – have developed parallel to the shoreline, enclosing shallow stretches of water. The eastern coast of the USA and the coast of Holland are lined by barrier islands and lagoons. The city of **Venice** is built on supporting piles sunk into a series of islands formed from silt and clay within a lagoon separated from the Adriatic Sea by the Lido and other strips of land.

Tie-breaker

⑥

Q: Which country has the longest coastline?
A: Canada. Its coastline is 243 792 km (151 485 miles) long, including the shores of its 52 455 islands. Stretched out, this would go six times around the Equator. It represents about a quarter of the world's total coastline.

Fjords ⑦

Fjords are **deep valleys** that were gouged out by glaciers during the last Ice Age, and then flooded when the ice melted. This process has produced the most heavily indented coastlines around the world, in places as far apart as Chile, Norway, Alaska and the South Island of New Zealand.

Islands 1

Core facts

◆ There are more than **half a million** islands in the world.

◆ Islands are classified as either oceanic or continental. **Oceanic** islands rise to the surface from the ocean floor. **Continental** islands occur where a rise in sea level has drowned a region of continental shelf, leaving higher areas of land separated from the mainland. Some islands, such as South Georgia in the South Atlantic, are a combination of the two.

◆ **Coral** islands, also known as atolls, consist of a near-continuous ring of coral reef enclosing a lagoon. They mostly occur on the flanks of extinct volcanoes.

TROPICAL PARADISE
The island of Tahaa in the Society Islands, French Polynesia, rises from a coral reef surrounded by a lagoon.

Oceanic islands – volcanic

Volcanic islands form where underwater volcanoes and activity at plate margins create a build-up of lava that rises above sea level. At **mid-ocean ridges**, where two oceanic plates are **moving apart**, volcanoes are large and are fed by a constant stream of magma. Iceland, the Azores and St Helena all rise from the Mid-Atlantic Ridge. Where two plates **converge** and one plunges under the other a string of islands, known as a **volcanic island arc**, forms parallel to the plate boundary. The Aleutians, Indonesia and the islands of Japan formed in this way.

A **hotspot** is a fixed magma source deep in the mantle below the moving crust. As the crust moves across the hotspot a chain of progressively older islands forms. The Hawaiian Islands formed in this way.

NEW-BORN ISLAND
Surtsey emerges off the coast of Iceland in 1963.

HOTSPOT As the ocean crust moves over the hotspot the first volcano is carried away and a new one forms over the hotspot.

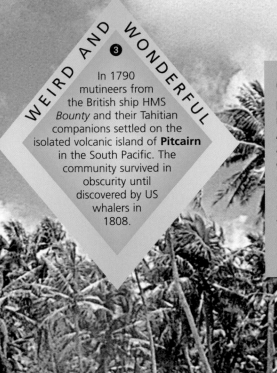

WEIRD AND WONDERFUL ❸

In 1790 mutineers from the British ship HMS *Bounty* and their Tahitian companions settled on the isolated volcanic island of **Pitcairn** in the South Pacific. The community survived in obscurity until discovered by US whalers in 1808.

Continental islands ❺

Many islands are fragments of neighbouring continents. In the **Atlantic**, the British Isles, Newfoundland and the Falkland Islands were once connected to their neighbouring mainlands; as were Madagascar, the Seychelles and Sri Lanka in the **Indian** Ocean. The **Pacific** Ocean is ringed with volcanic islands, but New Zealand, New Guinea, Taiwan and Indonesia are all continental islands.

Islands in legends ❻

Avalon The 'island of apples', ruled by the witch Morgan le Fay. The legendary King Arthur was taken there to be healed after his final battle.

Atlantis A 'lost' island in the Atlantic, west of Gibraltar. According to Plato, in prehistory this rich island was swallowed by the sea following an earthquake. It has been associated with Scandinavia, America and the Canary Islands.

Utopia An imaginary island invented by Sir Thomas More in his book *Utopia* (1516) where the inhabitants are governed by reason and live in perfect contentment. The word 'utopia' has come to mean 'a perfect place'.

Oceanic islands – coral ❹

Coral islands form where a **coral reef grows** in shallow water from a solid base, often the flanks of a volcano. Reef-building corals grow in water with an annual average temperature of 24°C (75°F), and occur mainly in the **tropical waters** of the Indian and Pacific oceans, and also in the Caribbean.

A reef begins forming in the sunlit upper waters around the flanks of a volcano. At this stage it is known as a fringing reef. As the oceanic crust slowly subsides, and the volcano with it, new coral builds upwards towards the surface, forming a barrier reef around the island. Eventually, the volcano becomes completely submerged, leaving just the reef, or atoll, around a lagoon.

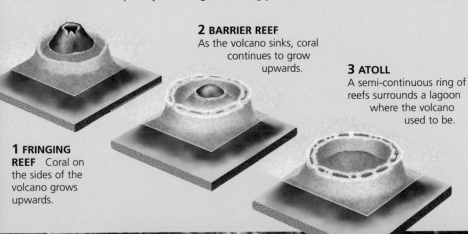

PERFECT ATOLL A lagoon has formed in the submerged crater of an old volcano.

1 FRINGING REEF Coral on the sides of the volcano grows upwards.

2 BARRIER REEF As the volcano sinks, coral continues to grow upwards.

3 ATOLL A semi-continuous ring of reefs surrounds a lagoon where the volcano used to be.

★ 21

Hawaii

There are **eight major** volcanic islands in the Hawaiian chain and many small coral islands. The crust of the western Pacific has moved westwards during the life of the hotspot, so the oldest islands are at the western end. The youngest island is Hawaii, at the eastern end, nearest to America's Pacific coast. Hawaii island currently has three active, one dormant and one submarine volcano.

Islands 2

Name origins ❶

Greenland Named in 982 by Norse navigator Eric the Red to attract settlers.

Easter Reached by Dutch navigators on Easter Sunday, 1722.

Alcatraz Spanish for 'pelican', which are numerous there.

Galapagos Spanish for the giant tortoises that live there, originally meaning 'scaly'.

Isle of Man From the Gaelic word for 'mountain'.

Juan Fernandez, Robinson Crusoe and Selkirk group First settled by the explorer Fernandez, whose livestock kept castaway Alexander Selkirk alive for five years until his rescue. Daniel Defoe based Robinson Crusoe on Selkirk.

Windward Island group from Grenada to Dominica blown by trade winds.

Solomon A Spanish explorer saw natives with gold ornaments and thought he had found the land of Ophir from where, according to the Old Testament, gold was brought to King Solomon.

ANCIENT FACES
Some of more than 600 stone figures on isolated Easter Island in the eastern Pacific. The statues are unlike any monuments found elsewhere in Oceania, but the more ancient figures on the island display some affinities with early monuments found in South America.

Island records ❷

Five largest	**Greenland** (2175600km²/839780sq miles); **New Guinea** (808510km²/312085sq miles); **Borneo** (757050km²/292220sq miles); **Madagascar** (587000km²/226660sq miles); **Sumatra** (524100km²/202300sq miles).
Largest in Europe	**Great Britain** (229870km²/88370sq miles). Occupies 1/1000 of the Earth's surface, but has 1/100 of its population.
Largest in Mediterranean	**Sicily** (25460km²/9830sq miles). One of the most densely populated in the Mediterranean, with over 5 million inhabitants.
Largest island group	**Indonesia** (13,700 islands). Together they cover about 2 million km² (772000sq miles).
Newest island	Appeared on June 6, 1995, in Tonga's **Ha'apai** group in the south-west Pacific.
Fastest to emerge	**Surtsey**, near Iceland, was born in 1963 when a volcanic eruption broke the surface of the North Atlantic.
Remotest	**Bouvet Island**, South Atlantic, 2400km (1490 miles) south-west of the Cape of Good Hope, 1600km (990 miles) north of Antarctica. Now Norwegian.
Most populous	**Java**, Indonesia, has a population of 118.7 million people in an area of 132,186km² (51037sq miles), or 898 people per km² (2326 per sq mile).

★ 26

Greek Islands

There are around 2000 Greek islands. Many are the tips of submerged mountain chains that extend out from the mainland. A few, such as Santorini, are entirely volcanic. The Ionian Islands are to the west of the mainland, Corfu being the largest. The Aegean Islands are to the east. Crete is the most southerly island.

Island kiss ❸

The 1957 movie *Island in the Sun* was set on a fictitious British-ruled Caribbean island. Starring Harry Belafonte and Joan Fontaine, it was seen as groundbreaking at the time – not least for a scene where the black Belafonte kisses the white Fontaine.

Island prisons ❹

Many prisons have been established on islands where rugged coasts and treacherous currents make escape difficult and dangerous. Some of the world's best-known and most influential prisoners have been held in them.

Robben Island South Africa	Nelson Mandela was held in a maximum security prison here between 1964 and 1982.
Elba Italy	Napoleon I was exiled here in 1814, after being forced to abdicate: 'I want from now on to live like a justice of the peace', he said.
Alcatraz California, USA	Gangster Al Capone served part of his 11-year sentence for tax evasion in the federal prison here.
Isle of Wight Southern England	King Charles I of England was held in Carisbrooke Castle during the Civil War. Parkhurst maximum security prison is now here.
Devil's Island French Guiana	French army officer Alfred Dreyfus was imprisoned in this penal colony for treason. Henri Charrière wrote about his own imprisonment and escape from here in his novel *Papillon*.

Unique worlds ❺

Plants, seeds and animals spread from land to islands by drifting in the current, rafting on tree logs, being carried by the wind or by birds. Once isolated from their parent population, island colonisers evolve independently in response to their new circumstances. New species emerge that are **endemic**, or specific, to that island. The finches and giant tortoises of the Galapagos Islands, the lemurs of Madagascar and the kiwis of New Zealand are all endemic species.

EVOLUTIONARY ISLANDS The isolated Galapagos Islands have a large percentage of endemic species.

Unusual landscapes

Core facts ❶

◆ Since the late 19th century unusual natural features, areas of wilderness and wildlife have been protected by the establishment of **national parks** and nature reserves.

◆ **Volcanic activity** has created some of the most colourful and extraordinary landscapes, including hot springs, coloured lakes and a soft volcanic rock that is easily eroded and shaped by the forces of water or wind.

◆ Many landscapes contain a kaleidoscope of colours – seen in rocks, lakes and soil – caused by the **minerals** present or the sediment they contain, or by the effects of heat or burial beneath the Earth's surface.

Protecting the natural world ❷

Yellowstone was the world's first national park, established in 1872. Since then, more than 3000 national parks have been created around the world.

National park	Special features
Yellowstone USA	The oldest in the world. The world's greatest geyser area.
Royal Australia	The second-oldest in the world, established in 1879.
Kruger South Africa	The oldest in Africa. Home to the rare white rhinoceros.
Corbett India	The oldest in India, established in 1935 in the Himalayan foothills. Home to the Indian tiger.
Wood Buffalo Canada	The largest national park in the world, and a refuge for the American buffalo.
Kakadu Australia	One of the world's largest national parks, Kakadu contains some of the earliest remains of human settlement.
Nahanni River Canada	A pristine wilderness area that was declared the first World Heritage Site in 1978.

GRAND PRISMATIC SPRING This multicoloured pool is one of thousands of hot springs and geysers in Yellowstone Park.

CHAMPAGNE POOL This geothermal pool, in the Waiotapu reserve on New Zealand's North Island, effervesces when sand is thrown into it.

KARST COUNTRY The River Li winds around the feet of China's extraordinary Guilin Hills.

The Guilin Hills

These hills in southern China tower over paddy fields and fish farms. The tallest, called Piled Festoon Hill, is 120 m (400 ft) high. The area is an extreme example of a type of limestone landscape known as **karst**, which is shaped by the dissolving power of groundwater. Water has removed large volumes of rock, leaving clusters of steep-sided, narrow hills riddled with caves and passageways. Other karst landscapes include the Kras Plateau in Slovenia, the Cockpit Country in Jamaica and the Mammoth Cave region of Kentucky (USA).

Pigmented rock

Rock colour varies according to the **minerals** it contains. Iron is the most common 'dye': combined with varying amounts of oxygen, it produces a range of colours from deep brownish red to pale yellow ochre. Copper gives green or blue rock; chromium and cadmium, orange rock; and manganese produces pink rock.

THE 'PAINT POTS' Erosion has revealed this aptly named area of colourful clays in Queensland, Australia.

★ 794

Petrified trees

Strewn across the desert landscape of **Arizona** are multicoloured, rock-like logs. They are the petrified remains of conifers that grew in the area over 200 million years ago when it was a flood plain. The rapid burial of fallen trees preserved them, and mineral deposits gradually replaced the wood.

STONE LOGS The wood has been replaced by quartz.

Conical hills

Cappadocia in central Turkey is known for its extraordinary rock shapes. The area is dominated by the extinct volcano Erciyas Dagi. Millions of years ago the volcano erupted, throwing out ash over a wide area. The ash hardened to form tuff or tufa, a soft, white rock that is easily eroded and washed away. Rain run-off has carved ravines and then eaten into their sides to form a maze of strange cones and pillars.

TURKISH DELIGHT Rock towers and cones fill the landscape, some with caves cut into them.

Oceans and seas 1

Core facts ❶

◆ The **oceans** consist of a continuous body of water covering 360 million km² (139 million sq miles), or 70 per cent, of the Earth's surface.
◆ **Seawater** is seldom still – waves, currents and tides all create constant movement.
◆ Most **waves** are caused by the wind blowing across the water surface. The exceptions are tsunamis (see page 122), which are caused by large displacements of water at sea.
◆ Large drifts of water, or **currents**, flow round the oceans in set patterns. There are two types of current: surface (warm) and deep (cold).
◆ Most parts of the world have two high and two low **tides** each day.

Waves

Waves derive their energy and motion from the **wind**. All waves have the same characteristics. The top is known as a crest, and crests are separated by troughs. Wave height is measured from trough to crest, and wavelength from crest to crest.

As waves approach the shore, the water becomes shallower and the waves get higher. Finally, the wave front collapses, or breaks. The turbulent water created by breakers is called surf.

Excluding tsunamis, the highest wave officially recorded was 34 m (112 ft) from trough to crest. It occurred druing a hurricane in the Pacific Ocean.

WAVE POWER
Waves consist of energy passing through water. The wave energy moves, not the water itself. Each water particle follows a circular path and returns to almost the same position.

Tides

Sir Isaac Newton was the first person to explain the tides – the regular rise and fall in ocean levels – by applying the law of gravity. Tides result from the **gravitational pull** exerted on the Earth by the Moon, and to a lesser extent the Sun.

The Moon's gravity pulls at the Earth, creating a bulge in the oceans on the side facing it; the centrifugal force produced by Earth's rotation causes water to pile up on the opposite side as well. The tidal bulges remain in place through the day, and the Earth rotates through them. So each place passes through both bulges in each 24-hour period, producing high and low tides twice in 24 hours. The effect of the Sun is less powerful than that of the Moon, but when the Sun and Moon are in alignment, their combined gravity creates the highest tides, or spring tides. When the Sun and Moon form a right angle in relation to Earth, the lowest tides, or neap tides, occur.

SPRING AND NEAP TIDES
The gravitational pull of the Sun and Moon causes exceptionally high – or spring – tides. When they pull in different directions, the lowest – neap – tides occur.

THE HIGHEST TIDE
The largest tidal range anywhere in the world occurs at the northern end of the Bay of Fundy, Nova Scotia. During spring tide conditions, the maximum tidal range reaches 17 m (56 ft).

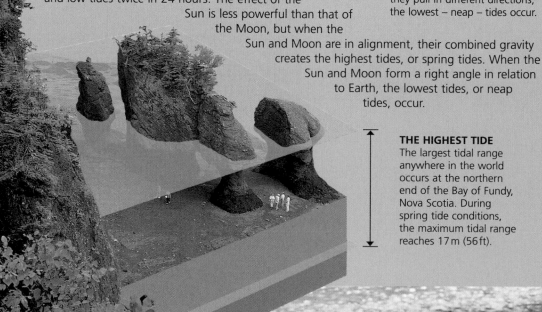

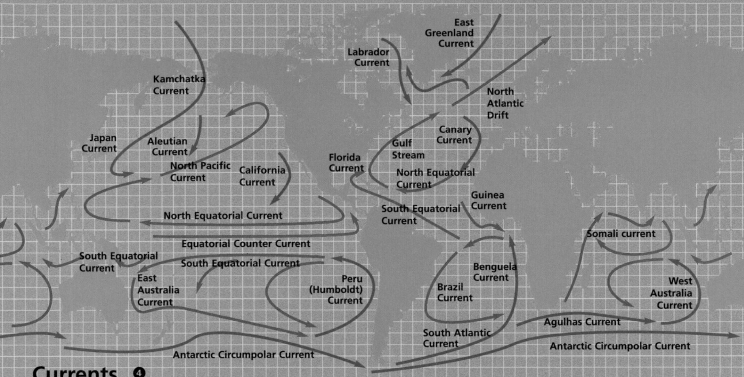

East
Greenland
Current

Labrador
Current

Kamchatka
Current

North
Atlantic
Drift

Canary
Current

Japan
Current

Aleutian
Current

Gulf
Stream

North Pacific
Current

California
Current

Florida
Current

North Equatorial
Current

Guinea
Current

North Equatorial Current

South Equatorial
Current

Equatorial Counter Current

Somali current

South Equatorial
Current

South Equatorial Current

East
Australia
Current

Peru
(Humboldt)
Current

Benguela
Current

Brazil
Current

West
Australia
Current

Agulhas Current

South Atlantic
Current

Antarctic Circumpolar Current

Antarctic Circumpolar Current

Currents ❹

Warm currents (red) transport warm water from the Tropics towards the Poles, and are driven by the wind. They follow roughly circular paths or gyres, circulating in a clockwise direction in the Northern Hemisphere and an anticlockwise direction in the Southern Hemisphere.

Cold, deep-water currents (blue) transport cold water from the Poles towards the Tropics. They are caused by decreases in temperature and increases in salinity, which make water more dense, so it sinks and spreads back towards the Equator.

Change of direction ❻

El Niño is a periodic reversal of Pacific Ocean currents. In a normal year currents flow west, driven by the trade winds. In an El Niño year, the trade winds fail and warm water drifts back eastwards towards Peru. El Niño events begin in December or January, hence its name, 'The Christ Child'. **La Niña** occurs when equatorial waters in the Pacific become colder than normal. Both affect global weather.

GREEN WATER In 1998 an algal bloom jetted across the Pacific, coinciding with upwellings of cold water at the start of La Niña.

Sea within a sea ❺

The **Sargasso Sea**, first described by Christopher Columbus in 1492, is a virtually stagnant area of warm water in the North Atlantic. It is the only sea entirely bounded by the Atlantic Ocean, being encircled by the Gulf Stream, the North Atlantic Drift, the Canary Current and the North Equatorial Current, which form a single, closed, clockwise-circulating system or gyre. Anything that floats is gradually drawn to the centre of the gyre. The surface is strewn with Sargassum seaweed (from the Portuguese *sargazo*, meaning weed), which is kept afloat by its berry-like bladders.

585

Whirlpools

The interaction between rising and falling tides can sometimes produce giant rotating currents, or whirlpools, with strong downdrafts. Famous whirlpools include Maelstrom off the coast of Norway and Charybdis between southern Italy and Sicily. A similar type of rotating current, known as a vortex, is produced where a deep, narrow channel occurs near a coast.

QUESTION NUMBER

The numbers or star following the answers refer to information boxes on the right.

ANSWERS

209   *Sea* – published in 1952

⭐ **253**   **D: Triangle** (see map, this page) ⭐

297   *South Pacific* ❸

303   Sea ❷

310   Bay ❷

316   Suez ❾

346   Water ❹

399   **Sweden and Denmark** – now linked by a bridge

484   **Pacific and Atlantic** ❾

485   **Botany Bay** – named for the wealth of plants there

526   **B: The English Channel** – literally means 'the sleeve'

545   **A balloon** – the *Double Eagle V*

583   **Mediterranean** ❶

608   **False** (the Persian Gulf is the warmest) ❹

623   *Titanic* – starring Leonardo DiCaprio, 1997

866   **Japan's Inland Sea** – the bridge was completed 1998

895   **Canals** ❾

898   **Seas** ❷

933   **Brown** – they made the crossing in 1919

937   **Yellow Sea** – named for sediment from Yellow River

971   **Pacific** ❶

Name origins

Aegean Named after Athenian king Aegeus, who jumped into the sea believing his son Theseus was dead.
Atlantic Named after either the legendary island of 'Atlantis', or the Greek giant 'Atlas', who bore the heavens on his shoulders.
Black Probably describes its appearance in bad weather.
Caribbean The Caribs were indigenous people of the island group from which the sea takes its name.
Mediterranean Latin for sea 'in the middle of the land', after its surroundings. ❶
Pacific Portuguese explorer Magellan called it the 'calm sea,' having encountered no storms on a journey across it.
Red Named after the algae that bloom in its waters, or the sandstone on its coasts.

What's the difference? ❷

Ocean One of five vast, deep, open expanses of saltwater: Pacific, Atlantic, Indian, Southern and Arctic.
Sea An expanse of saltwater partially bounded by land and shallower than an ocean.
Bay A wide inlet where the sea curves into the land.
Gulf A deep inlet with a narrow mouth, almost completely surrounded by land.
Strait A narrow strip of water connecting two large bodies of water.

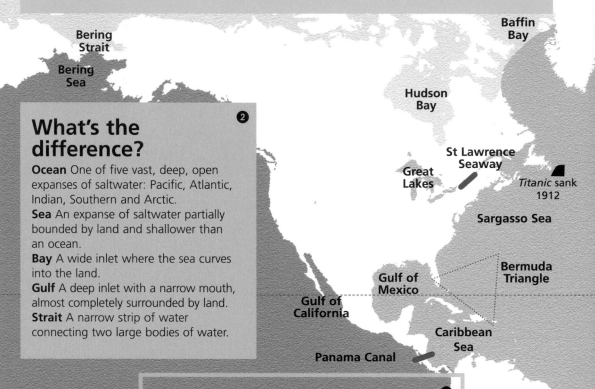

Pacific Ocean ❸

◆ Area 181 000 000 km² (70 000 000 sq miles)
◆ The **largest** ocean. Covers about one-third of the Earth's surface, and contains more than half the water on the planet.
◆ At its **widest point** the Pacific Ocean reaches almost half way round the world.

Sea facts and records ❹

Bering Sea	**Frozen** from October to June
Hudson Bay	The largest sea defined by **shoreline** length; **frozen** from November to July
Mediterranean Sea	Virtually **tideless**
Baltic Sea	Almost **freshwater** because of the large number of rivers that flow into it
Red Sea	The **saltiest** sea (some salt lakes, such as the Dead Sea, are even saltier)
Caspian Sea	The largest **inland** sea
Persian Gulf	The highest recorded surface water **temperature**
South China Sea	The **largest** sea defined by area

Arctic Ocean

- Area 14 056 000 km^2 (5 270 000 sq miles)
- The **smallest** and shallowest ocean.
- Contains about 1 per cent of the Earth's saltwater.
- A sheet of ice 4 m (13 ft) thick covers it for most of the year.

❺

Straits and canals **❾**

Bering Strait	Provided a land bridge between Asia and North America during last Ice Age.
Panama Canal	Opened in 1914.
St Lawrence Seaway	Links Atlantic with the Great Lakes.
Strait of Gibraltar	Provides the only natural access by sea to the Mediterranean.
Strait of Magellan	Separates mainland South America from Tierra del Fuego.
Suez Canal	Opened in 1869.

★ **253**

Mystery spot

The area known as the **Bermuda Triangle** is bounded by the coasts of Florida, Bermuda and Puerto Rico. It has been suggested that more than 50 ships and 20 aircraft have disappeared mysteriously in the area, but these figures are not borne out by scientific investigation.

Barents Sea

Baltic Sea

North Sea

English Channel ←

Bay of Biscay

Black Sea

Bosporus

Caspian Sea

Aral Sea

Mediterranean Sea

Suez Canal

Persian Gulf

Red Sea

↑ Strait of Gibraltar

Sea of Okhotsk

Sea of Japan

Yellow Sea

East China Sea

Tropic of Cancer

Philippine Sea

Atlantic Ocean **❻**

- Area 106 450 000 km^2 (41 100 000 sq miles)
- 150 million years old.
- More major rivers drain into the Atlantic than into any other ocean.
- The Atlantic and Pacific Oceans are linked by the Arctic Ocean in the north and Drake Passage in the south.

Arabian Sea

Bay of Bengal

Andaman Sea

South China Sea

Indian Ocean

- Area 73 556 000 km^2 (28 400 000 sq miles)
- Comprises about one-fifth of the total area covered by sea water.
- Formed around 140 million years ago. **❼**

Molucca Sea

Java Sea

Timor Sea

Gulf of Carpentaria

Equator

Coral Sea

Tropic of Capricorn

Botany Bay

Tasman Sea

Bass Strait

Southern Ocean **❽**

- Area 20 327 000 km^2 (7 850 000 sq miles)
- Officially designated the fifth ocean in 2000, the Southern Ocean includes all water lying south of latitude 60°S.
- In winter more than half the surface is covered by ice.

Ross Sea

Under the sea

Core facts ❶

◆ The **features** on the ocean floor are as varied as those of the continents, with canyons, volcanoes and mountain ranges.
◆ The **average depth** of the oceans is around 4000 m (13000 ft).
◆ **Mid-ocean ridges** occur where the seafloor is spreading, pushing continents apart as new material erupts from Earth's interior. In other places, oceanic plates push together, and ocean crust is returned to the interior.
◆ **Trenches** form where one plate sinks below another, dragging down the ocean floor.
◆ The ocean is divided into four **depth zones**: sunlit, twilight, abyssal and hadal.

Major underwater features ❷

The **mid-ocean ridges** occur in every ocean and form a linked system that runs for 80000 km (50000 miles). **Trenches** average 100 km (60 miles) wide and can be thousands of kilometres long.

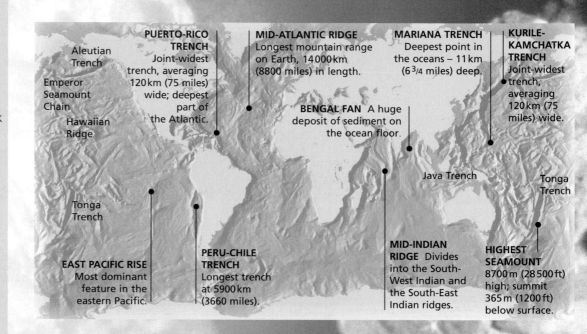

PUERTO-RICO TRENCH Joint-widest trench, averaging 120 km (75 miles) wide; deepest part of the Atlantic.

Aleutian Trench

Emperor Seamount Chain

Hawaiian Ridge

Tonga Trench

MID-ATLANTIC RIDGE Longest mountain range on Earth, 14000 km (8800 miles) in length.

MARIANA TRENCH Deepest point in the oceans – 11 km (6³/₄ miles) deep.

KURILE-KAMCHATKA TRENCH Joint-widest trench, averaging 120 km (75 miles) wide.

BENGAL FAN A huge deposit of sediment on the ocean floor.

Java Trench

Tonga Trench

EAST PACIFIC RISE Most dominant feature in the eastern Pacific.

PERU-CHILE TRENCH Longest trench at 5900 km (3660 miles).

MID-INDIAN RIDGE Divides into the South-West Indian and the South-East Indian ridges.

HIGHEST SEAMOUNT 8700 m (28500 ft) high; summit 365 m (1200 ft) below surface.

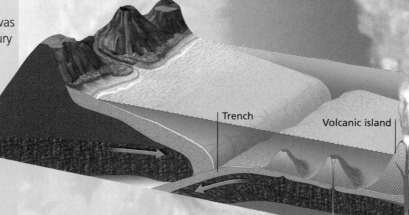

Trench

Volcanic island

Guyot: a flat-topped seamount

Mid-ocean ridge

Abyssal plain

The ocean basin ❸

The seabed is divided into two regions, the continental margins and the ocean floor, also known as the abyssal plain. The margins include the continental shelf, a shallow platform adjacent to land, and the continental slope, which falls away to the ocean floor. The floor consists of featureless plains dotted with underwater mountains known as seamounts, guyots and volcanic islands.

Sonar ❹

Sonar, short for Sound Navigation Ranging, is a method of detecting the **distance and location** of underwater objects. Sound waves sent out by an acoustic projector are reflected back by objects in the water and analysed to determine range, bearing and shape. Originally developed to detect icebergs, sonar was used to locate submarines during World War II.

Depth zones ❺

As ocean depth increases, light levels decrease and the temperature drops.

Sunlit zone: 0–200 m (650 ft) – Sunlight penetrates this region, enabling photosynthesis to take place. Microscopic plants provide food for large numbers of species.

Twilight zone: 200–1000 m (3000 ft) – Cold, and dimly lit.

Abyssal zone: 1000–5000 m (16 000 ft) – Reaches to the ocean floor. Temperature drops to around 5–6°C (41–43°F). The only light comes from signals produced by creatures to attract prey or mates.

Hadal zone Below 5000 m (16 000 ft) – also called the trench zone. Temperatures drop to around 1–2°C (34–36°F).

Sunlit zone

Twilight zone

Abyssal zone

Hadal zone

Seamount: a submerged volcano

Continental shelf

Continental slope

Saltwater ❻

The sea's **salinity** depends on the amount of salt and other substances dissolved in it. The major components are **sodium** and **chlorine**, which together form sodium chloride, or salt. These two elements, together with magnesium, potassium and calcium, make up 90 per cent of the elements dissolved in seawater.

Salinity affects the **density** of water, which affects the level at which a ship floats. Ships sit higher in cold, salty water than in warm or fresh water. The lines shown below, which appear on ships' hulls, show the lowest at which a fully loaded ship can safely ride in water with different temperatures and salinity.

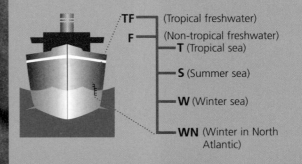

TF (Tropical freshwater)
F (Non-tropical freshwater)
T (Tropical sea)
S (Summer sea)
W (Winter sea)
WN (Winter in North Atlantic)

★ 449

The ocean floor

The ocean floor is covered by a thick layer of sediment that is a combination of marine plant and animal remains, grains of weathered rock from land erosion, and minerals crystallised out of the water. On deep ocean floors a 1 cm (1/2 in) layer can take anything from 500 to 50 000 years to form. In the Gulf of Mexico, where the Mississippi deposits its sediment, the ocean floor sediment is several kilometres thick.

BLACK SMOKER ❼
A hydrothermal vent, or 'black smoker', on the ocean floor pours out mineral-rich fluid. Hydrothermal vents occur on the flanks of mid-ocean ridges. This one, known as the Saracen's Head, is in the Mid-Atlantic.

QUESTION NUMBER

The numbers or star following the answers refer to information boxes on the right.

ANSWERS

- **69** **B: Volcanic crater** ❸
- **261** **Hot** ❶
- **265** **Hot** ❻
- **337** **Lava** ❷
- ★ **372** **Black** ★
- **643** **C: Valley of Hell** ❻
- **649** **D: A vent in a volcano** ❷
- **656** **Volcano** – in *Little Essays of Love and Virtue*, 1922
- **673** **Volcanoes** – different stages of a volcano's life
- **822** **Fire and metalworking** ❶
- **823** **Caldera** ❸
- **826** **Quiet period between eruptions**
- **910** **Beneath a volcano** ❷

Volcanoes 1

Core facts ❶

◆ The **word volcano** comes from Vulcan, the Roman god of fire. It applies to openings in the Earth's crust that spew out molten lava, ash, steam and gas, and to the cone of ash and lava that builds up around the opening.

◆ Most **volcanoes occur** along plate margins or at 'hotspots' in the Earth's crust.

◆ Volcano **shapes and types** are dependent on the type of lava and other material ejected and on the style of eruption.

◆ **Active** volcanoes are currently erupting or showing unrest. **Dormant** volcanoes are not erupting but may erupt again. **Extinct** volcanoes are no longer expected to erupt.

Anatomy of a volcano ❷

Deep beneath every volcano lies a reservoir of molten rock called a magma chamber. Stromboli's **magma chamber**, for instance, is 250km (155 miles) below the surface. Volcanoes erupt when the build-up of pressure in the chamber forces magma up a pipe, or **vent**, to the surface, where it erupts as **lava**. Solidified fragments of lava are ejected as clouds of ash. Successive flows of lava and ash build up a **volcanic cone**. Secondary cones occur at the exit of side vents. Vents where steam and gas emerge are called **fumeroles**. Solidified magma forms dykes, laccoliths and sills. Superheated ash clouds, or *nuées ardentes*, occur during particularly explosive eruptions.

★ 372

Obsidian

Obsidian is a dark, glassy type of rock that forms when viscous lava cools rapidly and solidifies without crystallising. Obsidian Cliff, in Yellowstone National Park, is formed entirely from obsidian. Opaque and slightly harder than window glass, obsidian has been used for weapons and cutting tools, ornaments, jewellery and, when polished, for mirrors.

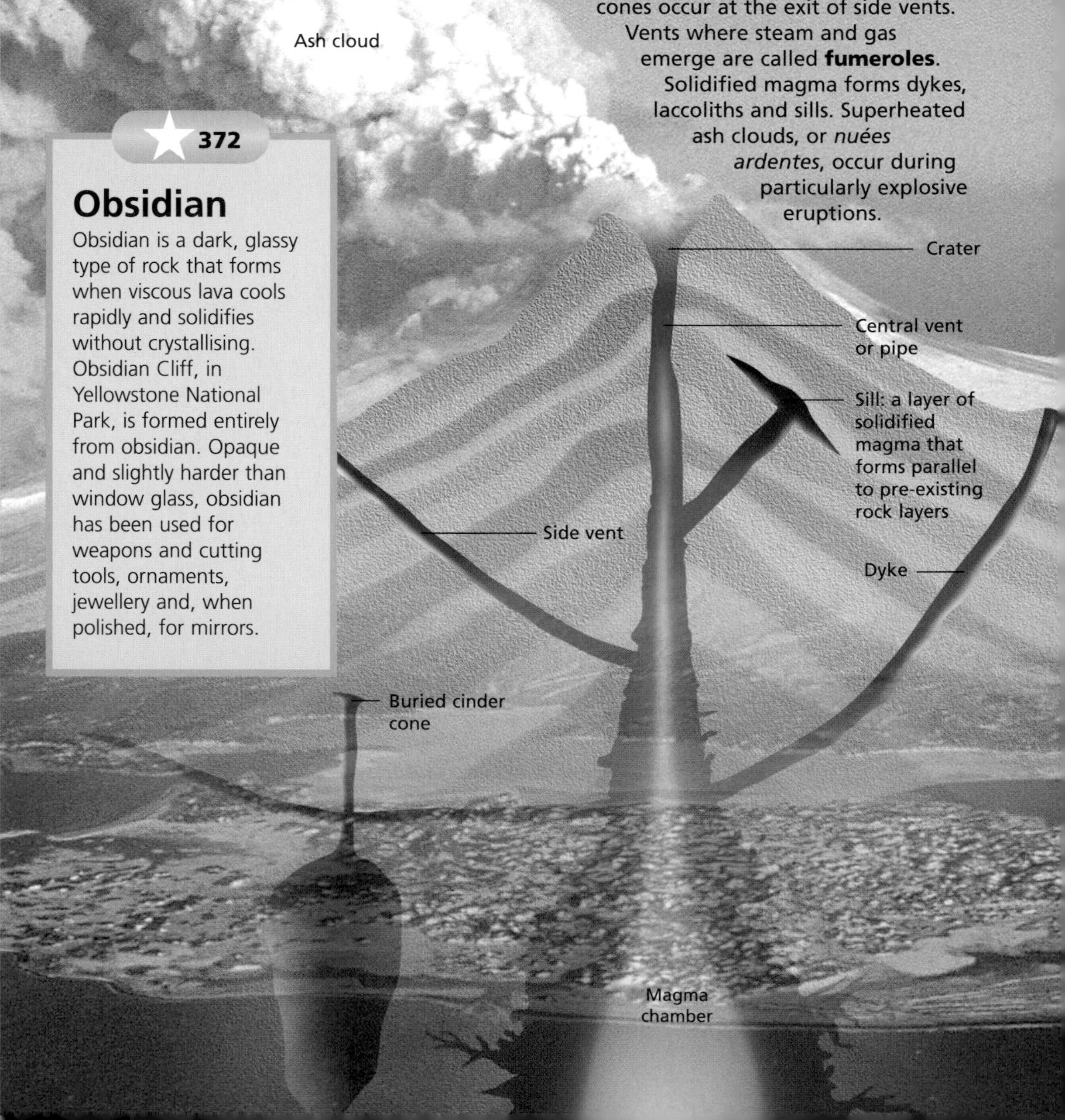

Ash cloud
Crater
Central vent or pipe
Sill: a layer of solidified magma that forms parallel to pre-existing rock layers
Side vent
Dyke
Buried cinder cone
Magma chamber

Volcano types

The **classic cone** shape associated with volcanoes is created by the build-up of successive layers of lava and ash erupting from a central vent. Not all volcanoes are this shape. The thin, runny lavas of Hawaii, for example, form very shallow-sided **shield** volcanoes, while runny lava erupting from a fissure rather than a vent forms stepped plains like the Deccan Traps of India. **Volcanic domes** are created by the slow release of thick lava. When volcanoes have been dormant or extinct for some while, the central part of the cone can collapse inwards, forming a large circular crater called a **caldera**.

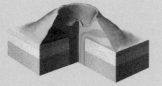   

Volcanic dome Built up by the slow release of thick lava.

Shield volcano Shallow-sided lava cone typical of Hawaii.

Composite volcano Classic, steep cone of lava and ash layers.

Caldera Collapsed cone. These craters are often filled by lakes.

Types of eruption

The way a volcano erupts depends on the composition of the lava and the amount of gas it contains. The thicker the lava or the higher the gas content, the more explosive the eruption.

Eruption types are named after typical examples – except for the Plinian, which is named after Pliny the Elder, who described the eruption of Vesuvius in 79 AD. Some volcanoes, such as Etna, erupt in different ways at different times, depending on the level of pressure and the lava content.

 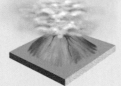 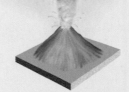 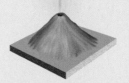 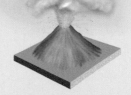

Peléean Thick, very gassy lava; explosive eruptions and gas clouds.

Hawaiian Runny lava, fire fountains; non-explosive.

Strombolian Runny lava with a high gas content. Small, frequent explosions.

Plinian Explosive eruptions; ash and rock thrown high into the air.

Vulcanian Rare, explosive eruptions throw ash and rock fragments long distances.

Mythology ❺

Volcanoes have often been associated with deities, such as **Ahriman**, the 'Destructive Spirit' of Zoroastrianism, who is shackled at the Earth's core. The Greeks thought of volcanoes as the workshops of **Hephaestus**, their god of fire, whose Roman counterpart was **Vulcan**. Another story says the giant **Typhon**, father of Cerberus, was imprisoned beneath Mount Etna and the Earth trembles when he turns.

Hot springs and geysers

Hot springs are created when groundwater is heated by the surrounding rocks, particularly in volcanic areas such as Iceland, Rotorua in New Zealand and Yellowstone Park. **Geysers** are fountains of boiling water that erupt under pressure.

There are 182 geysers in Yellowstone Park, including Old Faithful and Steamboat, currently the largest in the world with eruptions up to 90 m (300 ft) high. New Zealand holds the record for the world's highest-ever eruption of 500 m (1650 ft), in 1903, from a now-extinct geyser.

MAN-MADE ODDITY Fly Geyser in Nevada, USA, developed from artificial wells. Calcium in the well water has solidified to create the strange shapes.

116

QUESTION NUMBER

The numbers or star following the answers refer to information boxes on the right.

ANSWERS

53	**Volcanic eruption and tidal wave ❶**
232	*The Lost World* – novel and film, both 1997
237	1944 ❶
★ **314**	Etna ★ ❷
356	*Dante's Peak* – starring Pierce Brosnan, 1977
574	Mount St Helens ❺
578	Popocatépetl, Mexico – means 'smoking mountain'
579	Mt Erebus, Antarctica – has been active since 1972
580	Heimaey, Iceland – this eruption was in 1973
811	A volcano – fictional, but based on Cascade Range
821	*Los Angeles* – starring Tommy Lee Jones
824	*Journey to the Centre of the Earth* – publ. 1864
827	Edinburgh – extinct for over 350 million years
828	Pompeii and Herculaneum ❶ ❹
829	*Joe Versus the Volcano* – with Meg Ryan, 1990
929	Herculaneum ❶ ❹

Volcanoes 2

Famous volcanoes ❶

There are over 10 000 identified volcanoes in the world, 500 of which are active.

Highest volcano above seabed: **Mauna Loa** Hawaii		Although only some 4170 m (13 681 ft) above sea level, Mauna Loa, has its roots on the seabed. Measured from there it is not only the highest volcano in the world but also the highest mountain at 10 200 m (33 480 ft).
Highest active volcano above sea level: **Cotopaxi** Ecuador		Reaching a height of 5911 m (19 392 ft), Cotopaxi has erupted 50 times since 1738. The last recorded eruption occurred in 1904 and in 1975 the volcano awoke for a short time.
Most active currently: **Kilauea** Hawaii		Kilauea has been erupting lava continuously since 1983 at a rate of about 5 m^3 (174 cu ft) per second. The name Kilauea means 'spewing' or 'much spreading'.
Most active in history: **Stromboli** Italy		Continuous mild eruptions of Stromboli, on the arc of Italy's Aeolian Islands, have been recorded since Roman times, earning it the title of 'Lighthouse of the Mediterranean'.
Most people killed: **Tambora** Indonesia		The 1815 eruption of Tambora created an ash column 44 km (28 miles) high and killed 10 000 people directly. A further 82 000 died from starvation and disease.
Potentially the most dangerous: **Vesuvius** Italy		In AD 79 an eruption of Vesuvius destroyed the Roman towns of Herculaneum and Pompeii. A similar eruption today would potentially have a devastating effect on this densely populated region.
Greatest explosion: **Krakatau** Indonesia		Krakatau, in the Sunda Strait, exploded in 1883 destroying most of the island and creating a tsunami (tidal wave) that killed 36 000 people on neighbouring islands.

WEIRD AND WONDERFUL ❷

Between April and July 2000, **Mount Etna** produced up to 100 gas rings (smoke rings) a day, some perfect circles. Measuring up to 200 m (650 ft) in diameter, they rose to altitudes of about 5500 m (18 000 ft).

Tie-breaker ❸

Q: On which Caribbean island did Mont Pelée erupt on May 8, 1902, killing 29 000 people?
A: Martinique. White-hot ash rained down on the seaport of St Pierre, where only two people survived – one of them a prisoner in a well-protected cell in the jail.

Places and civilisations destroyed ❹

Volcanoes have wiped out entire towns, cities and even civilisations.

Crete	In 1626 BC the Greek island of **Santorini** exploded creating a 37 km (23 mile) high ash cloud and 50 m (160 ft) tsunami, which may have devastated the Minoan civilisation on nearby Crete.
Pompeii	Pompeii, 8 km (5 miles) downwind of **Vesuvius**, was buried with devastating speed under 3 m (10 ft) of ash from the volcano's AD 79 eruption. Poisonous sulphur fumes killed 2000 people.
San Juan	Although there was no loss of life, the eruptions of **Paricutin**, Mexico, in 1943–4 buried the towns of San Juan Parangaricutiro and Paricutin with lava. Only Paricutin's church can still be seen.
Goma	**Nyiragongo** in the D.R.Congo erupted in January 2002. Lava flows damaged 14 villages, making 12 500 families homeless. Around 500 000 people were evacuated from the city of Goma.

Mount St Helens ❺

On May 18, 1980, the nine-hour eruption of Mount St Helens in Washington, USA, removed the top 400 m (1300 ft) of the volcano's cone and thrust up a 19 km (12 mile) plume of ash that fell over 57 000 km² (22 000 sq miles). The blast felled trees and killed most wildlife and 57 people in a 550 km² (212 sq mile) area. Debris flows blocked the Columbia River. Volcanic dust was carried around the globe in the upper atmosphere.

VOLCANIC SCARS More than 20 years on, the vegetation and drainage in the area around Mount St Helens is still recovering.

★ 314

Mount Etna

The Sicilian volcano takes its name from the Phoenician *attuna* meaning 'furnace'. Lying on the boundary between the European and African continental plates, Etna is the highest active volcano in Europe.

Major eruptions occurred in 122 BC, AD 1669 (destroying the city of Catania), and 1787. Despite being one of the world's most active volcanoes, Etna has been responsible for only 73 known deaths.

LIGHT SHOW An explosive eruption rains down glowing lava bombs onto the flanks of Mount Etna.

RIVER OF FIRE Thick, runny, gas-rich lava spews from a rift on Kilauea, Hawaii, and flows slowly downhill.

Earthquakes

Core facts ❶

◆ An earthquake is any **motion of the ground**, no matter what the cause. They range from being imperceptible to being so violent that they can raze cities. No part of the world is exempt from earthquakes.
◆ Earthquakes happen as a result of the **release of energy** when rocks come under pressure, mainly from plate movements, shift or fracture.

◆ A few earthquakes are triggered by meteorite impacts, man-made explosions, or the weight of water, for instance behind a dam.
◆ Earthquakes produce three main types of **seismic waves**: primary (P), secondary (S) and surface (L).
◆ Earthquake **prediction**, the best defence against their effects, is still an inexact science.

Where earthquakes happen ❷

Most earthquakes occur at **plate boundaries**; 70 per cent occur around the Pacific, including the San Andreas fault region. Earthquakes also occur within continents, such as Tangshan in 1976. This map shows the regions with the greatest concentration of earthquake activity.

Mountain regions around the Mediterranean, through Turkey and Iran to the Himalayas.

The Pacific rim: Japan, the Philippines, western Alaska, California's San Andreas fault, Chile, and several island chains.

Ocean ridge system winding around the globe. Frequent low-intensity quakes.

What happens? ❸

When rocks can no longer absorb the build-up of pressure they break or shift along **fault lines**, releasing the pent-up energy as waves of vibrations radiating out from the centre, or **focus**, of the failure. We feel the surface waves as an earthquake, which can last just a few seconds or as long as four minutes. The **intensity** of the earthquake and the **damage** it wreaks are measured on the Richter and Modified Mercalli scales (see pages 152–3).

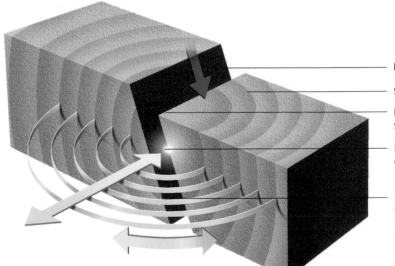

Fault line

Surface waves do the damage.

Epicentre The point on the surface directly above the focus.

Focus The centre of the earthquake.

Primary and secondary waves radiate out from the focus.

MEXICAN QUAKE In 1985 vibrations from a mild earthquake 400 km (250 miles) away were amplified by the rock on which Mexico City is built, causing extensive ground movement and damage.

Earthquake records ❹

Place	Date	Key fact
Shanshi China	1556	Highest known death toll – 830 000 people died. China has the worst record for earthquake deaths.
Lisbon Portugal	1755	Worst in Europe – 8.7 on Richter scale. 70 000 people died and the entire city was destroyed. Tremors were felt in North Africa, England and Luxembourg.
Yakatut Bay Alaska	1899	Greatest recorded vertical movement – the coast rose in places by 14.5 m ($47^1/2$ ft).
San Francisco USA	1906	Destroyed most of the city, and a huge fire burned for $3^1/2$ days. Survivors included the celebrities Enrico Caruso and John Barrymore.
Off Chile	1960	Largest earthquake ever recorded – registered 9.5 on the Richter scale.
Alaska	1964	Longest recorded – 4 minutes (Lisbon in 1755 was a 10-minute series of shocks).
Tangshan China	1976	Deadliest in the 20th century – 8.0 on the Richter scale. 655 000 people died.
Kobe Japan	1995	Worst damage financially – over US$147 billion of damage. 5500 people died.

Destructive power ❺

Place	Date	Feature
Rhodes Greece	c.225 BC	The Colossus, a huge bronze statue of the sun god Helios, one of the Seven Wonders of the World, was toppled.
Alexandria Egypt	AD 365	The lighthouse (Pharos), one of the Seven Wonders of the World, was toppled in an earthquake that killed 50 000 people.
Port Royal Jamaica	1692	This pirate haven was virtually destroyed – seen as punishment for the port's sinful reputation.
Unsen Japan	1793	The whole island and 53 000 inhabitants sank beneath the Satsuma Sea.
Messina Sicily	1908	The entire city was destroyed. Only one building survived.
Assisi Italy	1997	Earthquakes damaged the basilica of St Francis, destroying ceilings and frescoes. St Francis was said to have cured a boy who had been injured in an earthquake.

⭐ 104

Fore- and aftershocks

Foreshocks are minor precursors that can occur before a major earthquake, and are often used to predict a larger seismic event. **Aftershocks** are normally weaker shocks that happen days or weeks later. They can cause extensive damage to structures already weakened by the initial quake.

Earthquake gods ❻

In the Bible, **Yahweh** speaks to the prophet Elijah as an earthquake. In African myth earthquakes are caused by the dead of the underworld. The Greek god of earthquakes was **Poseidon**, also god of the sea. His Roman counterpart was **Neptune**. In Siberia the earthquake god is **Tuil**, who rides a sleigh beneath the Earth.

Landslides and avalanches

Core facts ❶

◆ **Landslides** – a general term covering all types of surface movement – and **avalanches** involve the mass movement downhill of soil, rock or snow under the action of gravity.
◆ **Vibrations** from earthquakes, explosions or sudden loud noises can destabilise snow or earth, resulting in avalanches and landslides.

◆ **Rainfall** or seeping groundwater can loosen rock or soil on a slope.
◆ Lahars are **mudflows** created during a volcanic eruption when water from a crater lake or melted snow mixes with ash.
◆ Avalanches are also triggered by the weight of extra snow falling on a sloping snowfield.

Landslides ❷

The **downhill movement** of rock, debris and soil under the pull of gravity is termed mass-wasting. It is divided into four types.
Fall The fastest type of movement, where loose material falls from a steep slope.

Slide Loose material slides en masse down a surface. The movement can be small or huge.
Flow Occurs when a mass of waterlogged, semi-fluid material moves rapidly downslope. Caused by rainfall, flooding, or melting snow.
Creep Soil slips downhill imperceptibly slowly.

ROCKFALL Rocks fall from a steepened or undercut surface. Their descent is by free-fall, bouncing or rolling.

ROCK SLIDE Happens on lower slopes. Slabs of rock break free and slide down a slope.

SLUMP Rock or soil slides a short way down a curved surface, creating an arc-shaped rupture.

CREEP The upper layers of soil slowly slip downhill, causing upright objects such as trees and fences to tilt.

KILLER MUD In 2000 a period of bad weather and heavy rainfall produced a slurry of mud and debris that swept through this Alpine valley in Slovenia.

DEBRIS/MUDFLOW The rapid movement of material, often started by high rainfall. Usually ends in a lobe or fan.

EARTH FLOW Drier, slower-moving flow. Often occurs at the base of a slump.

LAHAR Volcanic mudflows triggered by rainfall or melted ice. Move very fast down the mountainside.

UNDERWATER Slumps, flows and slides occur in sediments on underwater slopes. They often end in lobes or fans.

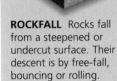

Worst-ever landslides and avalanches

Landslides and avalanches can be devastating in terms of human life and general destruction. Below are examples of some of the worst events of the past 100 years.

Location	Date	Fatalities	Comments
Gansu, China	1920	180000	Several landslides triggered by an earthquake.
Italian Alps	1916	10000	Series of avalanches over three days. 60,000 soldiers died in avalanches in World War I.
USSR	1949	12–20000	Earthquake triggered a landslide.
Mt Huascaran, Yungay, Peru	1970	40000	Earthquake shocks dislodged snow and rock on the slopes of the mountain.
Iran	1990	50000	Earthquake (7.7 on Richter Scale) caused avalanches that buried many villages.
Casita Volcano, Nicaragua	1998	1680	Mudflow caused by heavy rains from hurricane Mitch. Destroyed several towns.

★ 726
The speed of mud

Mudslides and flows are caused by unusually heavy rain or a sudden thaw. They consist mainly of mud and water, plus fragments of rock and other debris. They can reach 80 km (50 miles) long and 100 m (300 ft) high. Travelling at speeds of up to 55 km/h (35 mph) they can move houses off their foundations, or bury a place within minutes.

Snow avalanches

Most avalanches occur on slopes of 30–60°. Snow breaks free from the underlying surface and crashes downslope at speeds of 200 km/h (120 mph). There are two major types.

SLAB A large plate of compacted snow breaks off and moves downslope. Potentially devastating.

POWDER Loose surface snow slides downslope in a fan shape. Usually minimally destructive.

TOO CLOSE FOR COMFORT
The tips of an avalanche in the Swiss Alps reach chalets near the village of Evolene.

Disaster films

Earthquakes, volcanoes and storms have all been the subject of classic movies, including *The Perfect Storm* (2000), *Twister* (1996), *Earthquake* (1974), *Krakatoa, East of Java* (1969) and *San Francisco* (1936).

Floods and tsunamis

Core facts ❶

◆ **Coastal floods** are caused by bad weather, especially when it coincides with high tides, or by huge waves, tsunamis, that form at sea.
◆ **Rivers flood** when more water enters them than they can discharge. The usual causes are heavy rain or snow meltwater. Ice jams can temporarily dam a river, and release a torrent of water when they give way. Artificial causes include altered river courses, deforestation resulting in increased run-off, and burst dams.
◆ Sudden torrential rain concentrated in a narrow channel can produce a **flash flood**.

River floods ❷

Because so many people live near rivers, flooding is one of the most destructive geological hazards for humans, even in the 21st century.

River	Date	Comments
Ganges Bangladesh	1970, 1988	Deforestation upstream combined with storm surges in the delta area cause the most frequent catastrophic floods.
Huang He China	1931	Caused the greatest number of deaths in a single flood, estimated at 3.7 million.
Mississippi USA	1993	The USA's worst-ever flood lasted 144 days and covered over 2 375 000 km² (917 000 sq miles).
Arno Italy	1966	Inundated Florence, causing the greatest artistic loss of modern times – 3000 paintings were damaged and over 1 million books and manuscripts destroyed.

COMMUTERS People in the flooded Ganges delta make their way to work.

UNDER WATER A farm inundated in the Mississippi floods of 1993.

Flood mythology ❸

The universal Deluge that occurred around 2400 BC, described in Genesis, also appears in the Mesopotamian *Gilgamesh* epic, and in Greek, Chinese and Indian mythologies. The Sumerians recorded heavy rainfall in west and central Asia at a similar time.

⭐ **644**

Flash floods

Sudden, brief and fast-moving surges of water in sheets or along normally dry channels are typical of desert or semi-arid areas following intense, but short duration, rainfall. Flash floods can reach their **maximum flow** in minutes and achieve heights of 3 m (10 ft).

Tsunamis ❹

Tsunamis are catastrophically **destructive ocean waves** that occur when a submarine volcanic eruption, earthquake or sediment slump displaces a large volume of water. They can also be caused by a meteorite impact. Often undiscernible out at sea, tsunamis can reach **giant proportions** – up to 30 m (100 ft) high – when they reach shallow coastal waters. Tsunamis – the word comes from the Japanese for 'port' or 'harbour waves' – can travel at 800 km/h (500 mph) and frequently carry debris several miles inland.

SUDDEN SHIFT Millions of cubic tonnes of water can be displaced by a landslide, a downward displacement of the seafloor due to an earthquake, or lava from an underwater volcano.

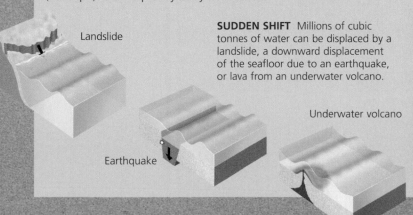

Landslide

Underwater volcano

Earthquake

Beneficial flooding ❺

The benefits of regular river flooding have been known since ancient times. In Egypt floodwaters provided **fertile silt and irrigation** for fields all along the Nile before the Aswan High Dam was built to provide water and electricity for the growing population. Now, silt is trapped behind the dam in Lake Nasser instead of fertilising the Nile valley. Downstream the river is flowing faster and cutting a deeper channel, which is affecting the efficiency of irrigation systems. And the Nile delta is deprived of its source of sediment and the coast is being eroded.

Tie-breaker

❻

Q: When people describe something as 'antediluvian', what do they mean?
A: That it is very old or old-fashioned. The word comes from the Latin for 'before the flood', and in its literal sense refers to the era before the Biblical Flood.

Time and seasons

Core facts ❶

◆ **Earth's axis** is tilted, so the angle at which the Sun's rays strike the Earth changes from month to month. This causes the changing seasons and variations in daylight length.
◆ During the **Northern Hemisphere summer** the Northern Hemisphere tilts towards the Sun and the Southern Hemisphere away from it.

During the **Northern Hemisphere winter,** the Northern Hemisphere tilts away from the Sun and the Southern Hemisphere towards it.
◆ According to the **astronomical calendar,** the seasons begin at specific points, known as the solstices and equinoxes, along Earth's orbit round the Sun.

Astronomical seasons ❷

When the northern end of Earth's axis tilts towards the Sun, the most direct sunlight falls on the Northern Hemisphere. The Sun reaches its highest angle in the sky there, giving the maximum number of **daylight hours** and the greatest solar heating. The North Pole has 24

hours of daylight. It is winter in the Southern Hemisphere, and a six-month **night** descends on the South Pole.

Six months later, the southern tip of the Earth's axis tilts towards the Sun and the situation is reversed.

JUNE The Sun is directly over the Tropic of Cancer. It reaches its highest point in the Northern Hemisphere sky and its lowest point in the Southern Hemisphere.

MARCH The Sun is overhead at the Equator. This point marks the beginning of the Northern Hemisphere spring and the Southern Hemisphere autumn.

SEPTEMBER The Sun is directly overhead at the Equator. This point marks the beginning of the Northern Hemisphere autumn and the Southern Hemisphere spring.

DECEMBER The Sun is directly over the Tropic of Capricorn. The most heat falls on the Southern Hemisphere, and it is winter in the Northern Hemisphere.

Solstice and equinox ❸

The **solstices** are the two days – June 21 and December 21 – when the Sun is directly over one or other Tropic, producing the maximum and minimum daylengths. On these two days the Equator is at its farthest point from the Sun.

The **equinoxes** (meaning day and night of equal length) occur when the Sun is directly overhead at the Equator, on March 21 and September 22.

THE FOUR SEASONS The yearly round of the seasons from early spring through to winter in New England, USA.

Counting seasons ❹

The greatest changes in heat and light occur in mid and high latitudes, and these regions have **four** seasons a year. Where the Sun is overhead, convection creates maximum storm activity. As this spot shifts between the two Tropics, heavy rainfall moves too, giving a cycle of **two** seasons, wet and dry. At the Equator, there is **one** season.

Leap years ❺

The Earth takes 365 days and 6 hours to orbit the Sun. This means that each year, the moment at which a new season begins (the moment the Sun is at its highest point) is six hours later than in the previous year. After four years the seasons are 24 hours behind, and this is corrected by adding an extra day, February 29, every fourth year.

GMT ⭐ 92

In 1884 the longitude of Greenwich, England, was chosen as 0°, or the Prime Meridian. Local time there, known as Greenwich Mean Time, became the basis for time zones round the world. GMT has been replaced by the more accurate Universal Time.

Time zones ❻

The world is divided into 24 time zones. Each zone represents 15° of longitude. Theoretically, the clock moves one hour forward for each zone to the east of the Prime Meridian, and one hour back for each zone to the west, although many variations occur.

The International Date Line is an imaginary line at 180° longitude. Cross it heading westwards and you gain a day; heading eastwards you lose a day.

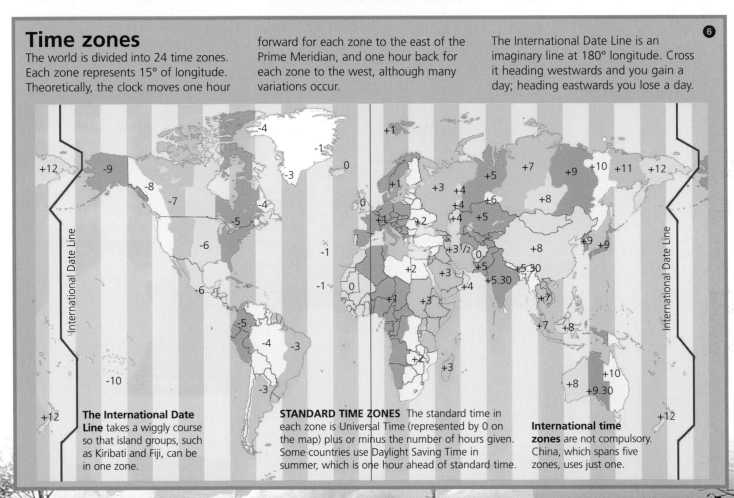

The International Date Line takes a wiggly course so that island groups, such as Kiribati and Fiji, can be in one zone.

STANDARD TIME ZONES The standard time in each zone is Universal Time (represented by 0 on the map) plus or minus the number of hours given. Some countries use Daylight Saving Time in summer, which is one hour ahead of standard time.

International time zones are not compulsory. China, which spans five zones, uses just one.

The numbers or star following the answers refer to information boxes on the right.

ANSWERS

6   **The Arctic** – it is better known as the polar bear

35   **Huskies** – can survive at -57°C (-71°F)

132   **D: Antarctic** ❹

247   **True** ❸

552   **Mediterranean** ❼

686   **Snow** – deepest snowfall ever recorded

702   **A temperate climate** ❻

798   **Climates** ❷

★ **872**   **True** ★

880   **True** – Verchojansk has recorded -67.7°C (-90°F)

960   **D: 320 km/h (199 mph)** ❹

991   **-89.2°C (-128.6°F)** – at the Russian scientific base, Antarctica

KEY TO REGIONS

Polar

Northern temperate

Mountain

Temperate

Mediterranean

Semi-arid (grassland)

Arid (desert)

Subtropical

Tropical

Coastal

Climate and vegetation 1

Core facts ❶

◆ **Climate** is the characteristic, day-to-day weather in a region based on a minimum of 30 years of recordings.
◆ A place's climate depends primarily on its **latitude**, as this determines how much heat it receives from the Sun and how much the seasons differ.
◆ **Other factors** that affect a region's climate include whether the prevailing air masses are hot or cold and moist or dry, and the distribution of land, oceans and mountains.
◆ Weather **extremes** are measured at weather stations. In order for a weather extreme to be declared a record, the station must have made a series of climatological observations over several years.

Climate types ❷

Climates fall into three main groupings, based on latitude.
◆ **Tropical** climates occur in low latitudes, and include desert regions near the Tropics as well as equatorial forests. They are dominated by equatorial air masses all year.
◆ **Temperate**, mid-latitude climates occur between the Tropics and the Arctic and Antarctic circles. Sub-tropical and sub-polar air masses alternate.
◆ **Polar** climates occur in high latitudes, between the Arctic and Antarctic Circles and the Poles. They are dominated by sub-polar and polar air masses.
◆ In addition, **mountains** and **coastal** areas have climates that differ from adjacent regions.

GREATEST MEASURED ANNUAL SNOWFALL
31 102 mm (1224 1/2 in) at Paradise, Washington state, USA, from February 19, 1971 to February 18, 1972.

Climate variations ❸

Latitude is the main factor affecting a place's climate, yet many areas on the same latitude have very different climates. For example, Mount Everest, the Sinai desert and the Florida jungle all lie approximately 28° north of the Equator. Everest has a cold, dry climate because of its altitude. The Sinai desert has a hot, arid climate. Florida is in an area bordered by warm ocean currents where the trade winds bring rain in summer, giving it a humid subtropical climate.

DRIEST PLACE
Atacama Desert, Chile.

Polar ❹

◆ Close to the poles the air is **cold** and **dry**.
◆ Winters are **long** and **cold**, lasting about 8–9 months; summers are slightly warmer.
◆ Precipitation takes the form of **snow**.
◆ **Antarctica** has the coldest, driest and windiest climate in the world. Average winter temperatures there are -55° to -60°C (-67° to -76°F); winds can reach 320 km/h (199 mph).
◆ Summer temperatures can reach 10°C (50°F).

★ **872**

Microclimates

Local variations within the climate of an area can be caused by ocean currents, forests and cities. The north-west coast of Scotland is warmed by the Gulf Stream, and has a mild climate where subtropical plants flourish. Towns and cities form heat islands. A microclimate can even exist under a stone.

Northern temperate ❺

◆ Similar to temperate zones, but winters can last up to nine months, and snowfall is heavy. Summers are short and warm.
◆ The average temperatures are -3°C (27°F) in winter, and over 10°C (50°F) in summer.

Temperate ❻

◆ Cold polar winds meet warm air, producing mild, changeable weather.
◆ There are four distinct seasons, with warm summers and cold winters with snow.
◆ Rainfall throughout the year.
◆ Temperatures vary widely with location.

Mountain ❽

◆ Temperatures drop with height, and rain and snow increase, creating vertical zoning.
◆ Zones can extend from tropical at the base to polar at the summit. Even at the Equator summits can be snow-covered due to low temperatures at the top.

Coastal ❿

◆ The sea gains and loses heat more slowly than land, and seasonal extremes occur about two weeks later than in adjacent inland areas.
◆ These regions have a narrower temperature range than nearby inland areas.

Arctic Circle

Subtropical ⓫

◆ Has a greater temperature range than tropical areas, and temperatures are generally lower, although still high.
◆ Has two seasons, wet and dry, which are similar in length, lasting about five to seven months each.

HIGHEST ANNUAL AVERAGE RAINFALL 11874.5mm (467$^{1}/_{2}$in), at Cherrapunji, Meghalaya state, India.

Tropic of Cancer

Tropical ⓬

◆ The air is always hot and moist, giving high temperatures and humidity all year. Rainfall is heavy for all but one or two months. Humidity is 90–100 per cent.
◆ Temperatures average about 26°C (79°F), with an annual fluctuation of less than 3°C (5°F).
◆ Rainfall can average 250cm (100in) a year.

Equator

HIGHEST 24-HOUR RAINFALL 1869.9mm (73$^{1}/_{2}$in), island of La Réunion, Indian Ocean, on March 15-16, 1952.

Tropic of Capricorn

HOTTEST AVERAGE ANNUAL TEMPERATURE 34.4°C (94°F) at Dalol, Ethiopia.

HOTTEST PLACE EVER RECORDED 57.8°C (136°F) at Al'Aziziyah, Libya, on September 13, 1922.

COLDEST PLACE EVER RECORDED -89.2°C (-128.6°F), at Vostok station, Antarctica, July 21, 1983.

Antarctic Circle

COLDEST ANNUAL AVERAGE -57.8°C (-72°F), at the Pole of Inaccessibility, Antarctica.

Mediterranean ❼

◆ Occurs around the Mediterranean Sea and on the western flanks of continents at similar latitudes north and south of the Equator – in California, South Africa and southern Australia.
◆ Has four seasons. Summers are hot and dry, with little rainfall for four to six months. Winters are cool and wet.

Semi-arid ❾

◆ These areas have a less extreme climate than arid regions, with a short wet season and smaller fluctuations in temperature between summer and winter.
◆ Strong winds and severe droughts are common.
◆ Rainfall is low, averaging around 250–760mm (10-30in) per year.

Arid ⓭

◆ Warm, dry air produces cloudless skies and little rain. Moisture evaporates quickly.
◆ Large temperature fluctuations occur between night and day and summer and winter.
◆ Average annual rainfall is less than 250mm (10in).
◆ Daytime temperatures average around 38°C (100°F).
◆ Night-time temperatures can drop below freezing.

QUESTION NUMBER

The numbers or star following the answers refer to information boxes on the right.

ANSWERS

- 3 **Black Forest gâteau** – the cherries grew in the valleys
- 47 **South America** ❹
- 49 **Asia** ❸
- 175 **Tundra** ❷
- 205 ***Little House on the Prairie* (1935)** ❹
- 208 **India** – book published 1894; Disney film 1967
- 252 **C: Canopy** ❾
- 308 **Plain** ❹
- 560 **Kilimanjaro** ❽
- 588 **Igloo** – traditionally built from blocks of ice
- ★ 590 **Tundra** ★
- 601 **True** ❼
- 639 **Purple Heart** – the trees are of the genus *Peltogyne*
- 685 **Amazon rain forest** ❾
- 784 **Prairie** ❹

Climate and vegetation 2

Vegetation types ❶

Vegetation is influenced by temperature, rainfall and humidity. Belts of characteristic vegetation have evolved in response to the prevailing climatic conditions.

In some regions, especially the mid-latitude temperate zones, much of the original natural vegetation has been cleared to make room for crop-growing.

Poles and tundra ❷

Between the Arctic ice cap and the taiga of the northern temperate zone is a region of **treeless plains**, or tundra. Below the topsoil the ground is frozen to a depth of about 5 m (16 ft). This frozen soil or **permafrost** supports plants such as mosses and lichens, which have adapted to cold conditions.

FROZEN WASTE Antarctica is bound in by ice. The Antarctic ice cap supports no vegetation.

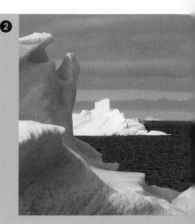

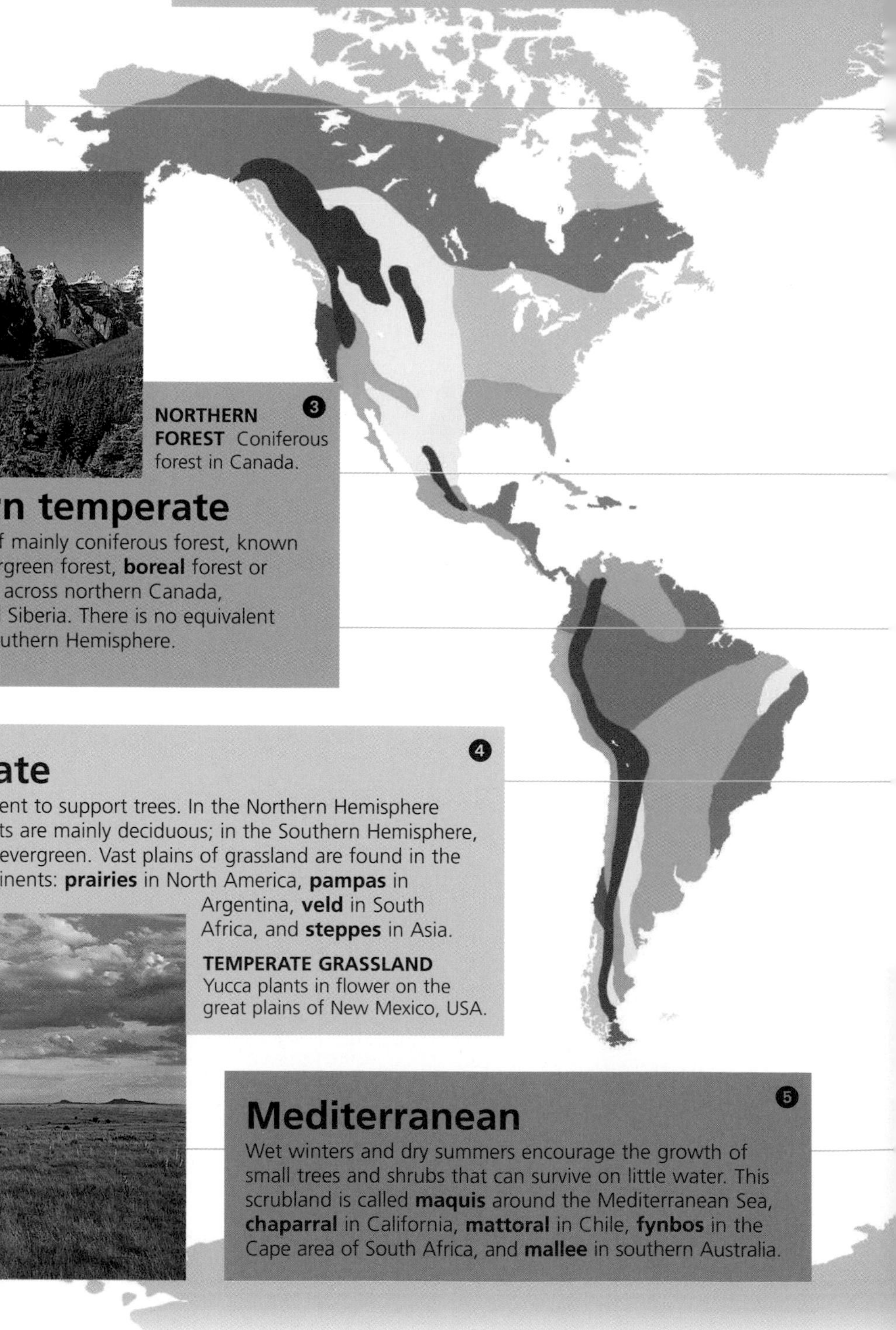

NORTHERN FOREST Coniferous forest in Canada. ❸

Northern temperate

A broken belt of mainly coniferous forest, known as northern evergreen forest, **boreal** forest or **taiga**, stretches across northern Canada, Scandinavia and Siberia. There is no equivalent region in the Southern Hemisphere.

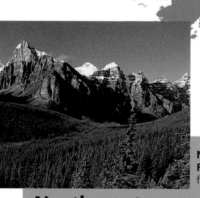

Temperate ❹

Rainfall is sufficient to support trees. In the Northern Hemisphere temperate forests are mainly deciduous; in the Southern Hemisphere, they are mainly evergreen. Vast plains of grassland are found in the interiors of continents: **prairies** in North America, **pampas** in Argentina, **veld** in South Africa, and **steppes** in Asia.

TEMPERATE GRASSLAND Yucca plants in flower on the great plains of New Mexico, USA.

KEY TO REGIONS

- Poles and tundra
- Northern temperate
- Mountain
- Temperate
- Mediterranean
- Semi-arid (grassland)
- Arid (desert)
- Subtropical
- Tropical

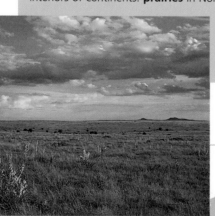

Mediterranean ❺

Wet winters and dry summers encourage the growth of small trees and shrubs that can survive on little water. This scrubland is called **maquis** around the Mediterranean Sea, **chaparral** in California, **mattoral** in Chile, **fynbos** in the Cape area of South Africa, and **mallee** in southern Australia.

PATTERNED GROUND The outlines of polygonals are marked by channels of rocks.

★ 590

Tundra patterns

In tundra regions the ground freezes for most of the year. In summer the top layer melts, but the water cannot drain because the ground below is still frozen. The continuous expansion and contraction of this waterlogged soil as it freezes and unfreezes creates ice wedges that split the ground into polygonal shapes. Water in underground springs may also freeze and force up hills of ice known as pingoes.

Mountain ⑧

The number of different zones depend on the mountain's latitude, but can range from **tropical rain forest**, in the case of the Andes, or **savannah** in the case of Kilimanjaro and Mount Kenya in East Africa, at the base, through **alpine** vegetation higher up, to glaciers and a snow-capped summit.

Arctic Circle

Tropical ⑨

A belt of **tropical rain forest** is centred on the Equator. Constant light, warmth and rainfall all year encourages continuous plant growth, creating the canopy of evergreen vegetation found in forests of the Amazon, Zaire and South-east Asia.

Tropic of Cancer

Equator

Semi-arid ⑥

These regions form in the interiors of continents and round the fringes of deserts, where rainfall is light. They are dominated by dry grassland, or **savannah**, and occur in North and South America, Africa, Asia, and Australia.

Tropic of Capricorn

CLOUD FOREST Pockets of cloud sit on the canopy of this Costa Rica forest. ⑩

Subtropical

Regions with a dry season and a very wet season develop tropical **monsoon forests**. In mountain areas the air is cool and condenses into cloud, fog and mist and **cloud forests** develop that have mixed tropical and temperate tree species.

Both types of forest suport fewer plant and animal species than rain forests, but more than temperate forests.

Antarctic Circle

Arid ⑦

What little rain deserts receive each year tends to fall all at once, leaving the land dry for the rest of the year. This lack of water, combined with drying winds, means that plants have to be efficient at collecting and storing water, like the **cactus**, to survive. Some, like the **sagebrush**, lie dormant for long periods.

DRY REGION The Atacama desert in the Andes foothills has sparse, arid adapted vegetation.

Earth's atmosphere

Core facts ❶

◆ Earth is surrounded by **gassy layers**, collectively known as the atmosphere, held in place by gravity. The atmosphere is very thin in relation to Earth's diameter, only extending about 600 km (360 miles) from the surface.
◆ The **five layers** of the atmosphere are defined by their temperature, which fluctuates from layer to layer.

◆ Significant amounts of **oxygen** in the atmosphere sustain most life on Earth.
◆ The atmosphere filters out much of the Sun's harmful **ultraviolet** radiation and burns up objects approaching from space, preventing most of them from reaching Earth.
◆ Ninety-nine per cent of atmospheric gases are concentrated in the bottom 80 km (50 miles).

Gases ❷

The atmosphere contains a mixture of gases. The lowest two layers consist of 78 per cent nitrogen, 21 per cent oxygen and 1 per cent other gases – including carbon dioxide, argon, methane, ozone, hydrogen and helium. Water vapour fluctuates between 1–4 per cent in the lower layers. The upper layers become gradually more rarefied, consisting mainly of widely spaced hydrogen and helium atoms.

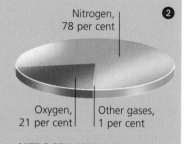

Nitrogen, 78 per cent

Oxygen, 21 per cent

Other gases, 1 per cent

NITROGEN ATMOSPHERE
More than three-quarters of the Earth's atmosphere is composed of nitrogen.

Altitude records ❸

Parachute jump	1960	Joe Kittinger jumped from a balloon at 30 840 m (102 800 ft). He broke the sound barrier falling at over 1120 km (700 mph), and survived.
Highest manned balloon	1961	Malcolm Ross and Vic Prather reached 34 000 m (113 740 ft). Prather drowned in the Gulf of Mexico on landing.
Unmanned solar aircraft	2001	Solar-powered *Helios* flying wing, built by NASA, flew to 29 413 m (94 000 ft) over the Pacific Ocean.
Manned single-engine aircraft	1938	Mario Pezzi flew a Caproni 161 single piston-engined aircraft to 17 083 m (56 046 ft).

HOT, COLD, HOT
Temperatures rise and fall from layer to layer of the atmosphere. They are hottest on the outer edge of the thermosphere, and coldest on the boundary of the thermosphere and mesosphere.

Atmospheric pressure ❹

The atmosphere exerts a downward force, also known as **air** or **barometric** pressure. At sea level it is 1 kg per cm² (14³/4 lb per sq in). Because air molecules are drawn to Earth by gravity, pressure is greatest at the surface and decreases with altitude. Air pressure is measured in millibars (or hectopascals). Fluctuations, detected by barometers, reflect imminent changes in the weather. Falls indicate unsettled weather while rises suggest fair conditions.

⭐ 390

Hot and cold

The thermosphere is known as the heat sphere, or warm layer, because of its very high temperature, yet it would actually feel cold. The temperatures recorded there are due to kinetic energy, not stored heat. There are too few molecules in the thin air to transfer enough heat for any difference in temperature to be felt.

SPACE TELESCOPE
The Hubble Space Telescope orbits Earth in the exosphere, the upper reaches of the atmosphere.

AURORAE The Northern and Southern Lights are seen in higher latitudes. They occur when charged solar particles enter the thermosphere.

METEOR SHOWERS
These trails of space debris burn up in the mesosphere.

BALLOONS
Specially designed balloons have risen more than 30 km (18 miles) and into the stratosphere.

CONVENTIONAL AIRCRAFT
These fly in the troposphere.

EXOSPHERE This outermost layer continues until it merges with the interplanetary gases of space. Temperatures range from 300°C (570°F) to 1700°C (3000°F). Helium and hydrogen are the main elements, but molecules are widely spaced and the lack of gravity allows gases to escape into space. Shortwave radio waves are reflected back to Earth.

THERMOSPHERE This layer reaches to 600 km (370 miles) above sea level. It contains a region called the ionosphere, where atoms and molecules become electrically charged (ionised) by solar radiation, which generates heat. Temperatures rise dramatically, reaching 1700°C (3000°F) on the outer edge of the thermosphere. Some radio transmissions from Earth are reflected back.

MESOSPHERE This extends to 80 km (50 miles) above sea level. Temperature decreases with altitude to -100°C (-148°F) at the top of this layer.

STRATOSPHERE ❺ This reaches to 50 km (31 miles) above sea level. Temperatures can drop below freezing. The ozone layer is located in the stratosphere, about 24 km (15 miles) above sea level. Volcanic dust and some industrial chemicals, such as CFCs, find their way into this layer and damage the ozone layer. Concorde reaches altitudes of 18 km (11 miles) above the Earth.

TROPOSPHERE ❻ This extends from sea level to 8–16 km (5–10 miles) above the Earth's surface. The height varies with season and latitude – highest at the Equator and lowest over the Poles. Temperature falls with altitude, on average by 7°C per km (4°F per 1000 ft), down to a minimum of -58°C (-70°F). Ninety-nine per cent of weather happens in this layer. Pollen, dust, material ejected by volcanoes and meteoric fragments are found here.

Why is the sky blue? ❼

Sunlight is scattered in different directions by dust particles and gas molecules in the atmosphere. The shorter light waves at the blue end of the spectrum are scattered more effectively than the longer waves at the red end. Large amounts of blue rays and small amounts of the other colours are scattered across the sky. We see these light rays against the dark background of outer space, and so the sky appears to be blue.

World winds

Core facts ❶

◆ Winds are created by **differences in air pressure**. Hot air rises, creating a low-pressure area; cold air sinks, creating a high-pressure area. Air is pushed out of high-pressure areas towards low-pressure areas.

◆ The **movement of air** from high-pressure to low-pressure areas follows a basic worldwide pattern. Winds blow from the Equator towards the Poles, or from the Poles towards the Equator, but Earth's rotation deflects these sideways, a phenomenon known as the **Coriolis effect**.

◆ **Wind direction** is the direction from which the wind blows. The prevailing wind is the most frequent wind direction in an area.

◆ **Monsoons** are sudden seasonal reversals in wind direction. They bring heavy rain.

Coriolis effect ❷

Hot air rises in the Tropics and spreads north and south towards the Poles. Because of the Earth's rotation, the air does not travel in a straight line but is deflected – to the right (clockwise) in the Northern Hemisphere, to the left (anti-clockwise) in the Southern Hemisphere. This is called the Coriolis effect, after Frenchman Gustave-Gaspard de Coriolis (1792–1843) who first described the phenomenon.

The Coriolis force also affects the rotation of ocean currents. It increases in force from the Equator to the Poles.

SAND STORM One of the largest sand storms ever recorded swept 1600 km (1000 miles) across the eastern Atlantic from the Sahara in February 2000, proof of the awesome power of upper-level winds.

Global wind patterns ❸

Each hemisphere has **three wind systems**: tropical trade winds, mid-latitude westerlies, and polar easterlies. Hot air rises at the Equator and travels north and south. When it reaches latitude 30° north and south, known as the **horse latitudes**, it sinks, creating high-pressure areas. Some of this air travels back towards the Equator, and these air flows are known as the **trade winds**. They die out at the Equator, creating an area of calm, light winds known as the **doldrums**. The **westerlies** are eastward-moving winds that blow from the horse latitudes towards the Poles. At around 60° north and south they meet the **polar easterlies**, cold winds that blow out from high-pressure areas at the Poles. The westerlies and polar easterlies meet at the **polar front**. The difference in temperature causes warmer air to rise, and most of it goes back towards the Equator.

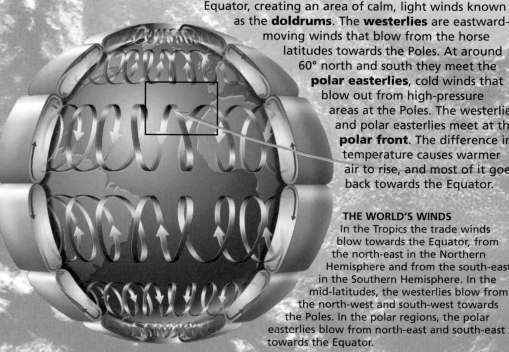

THE WORLD'S WINDS
In the Tropics the trade winds blow towards the Equator, from the north-east in the Northern Hemisphere and from the south-east in the Southern Hemisphere. In the mid-latitudes, the westerlies blow from the north-west and south-west towards the Poles. In the polar regions, the polar easterlies blow from north-east and south-east towards the Equator.

Monsoons

4

The strongest monsoons occur in southern Asia, northern Australia and Africa.

◆ In **winter** air over Siberia cools and sinks, causing a high-pressure area north of the Himalayas. This in turn creates winds that blow across India and out over the Indian Ocean.

◆ In **summer** the high pressure over Siberia weakens and a low-pressure area takes its place. This draws in moisture-laden air from the Indian Ocean, but its progress north is blocked by the Himalayas, producing heavy rain over the Indian subcontinent.

WATER WORLD Floodwaters from the seasonal Indian monsoon inundate a river basin and the surrounding field systems.

WINTER Cold, dry air descends north of the Himalayas. Warm air rises over the Indian Ocean and draws in colder air from the land.

SUMMER Warm air rises, creating a low-pressure area north of the Himalayas, and moisture-laden air is drawn in from the Indian Ocean.

958

Jet streams

Meandering bands of fast-moving westerlies, called jet streams, develop high in the atmosphere above polar and sub-tropical areas, at heights of 10-15 km (6-9 miles), where large differences in temperature and air pressure occur. They can reach speeds of 300 km/h (200 mph), but average about 130 km/h (80 mph) in summer and 70 km/h (40 mph) in winter.

Myths and legends

5

In many myths the winds are personified and often deified, reflecting their importance in agricultural and seafaring societies. The Ancient Greeks had **Boreas** (North Wind) and **Zephyrus** (West Wind), ruled by the minor god **Aeolus**. To the Sumerians, the wind was the breath of **Enlil**. The Aztec wind god **Ehecatl** was worshipped in round buildings, which offered no sharp corners to the wind. Other wind deities include Celtic goddess **Dana** and Mayan god **Huracan** (from which we get hurricane).

Local winds

Core facts ❶

◆ **Local differences** in temperature and air pressure create small-scale, regional or localised winds. These variations occur particularly in coastal regions and mountain ranges, or because of intense heating in inland desert areas.

◆ **Sea breezes** blow on warm days; **land breezes** on clear nights.

◆ **Mountain ranges** create high-pressure areas on their leeward sides, while mountain valleys experience uneven heating between the valley floor and sides.

◆ **Valleys** can funnel and intensify winds that blow down them.

◆ The fast-rising air currents known as **thermals** are, in effect, vertical winds.

Local wind types ❷

Specific local physical conditions create particular wind types.

SEA BREEZE Warm air over land rises. Cool wind blows in from sea.

LAND BREEZE Warm air over sea rises. Air over land cools and sinks, and blows out to sea.

FÖHN/CHINOOK-TYPE Starts warm, cools as it rises, and warms up again as it flows downhill.

MISTRAL-TYPE A cold downslope wind funnelled through a valley.

ANABATIC An upslope wind caused by the Sun heating the air.

KATABATIC Cold, heavy air moves down a slope.

CHINOOK CLOUDS Sunset lends these chinook wind-clouds a rosy glow over Banff National Park, Canada.

VALLEY MIST An early-morning mist forms in the base of a valley due to anabatic wind movements.

Whirling winds

Tornadoes are rapidly rotating columns of air that extend down from the base of a thundercloud to the ground. They range from several metres to several hundred metres in diameter. Some last for a number of hours and can travel hundreds of miles. **Waterspouts** are tornadoes that occur over water. They do not require a storm in order to form. **Dust devils** are smaller whirlwinds that occur over dusty ground.

GIANT WATERSPOUT ❻
A tornado-at-sea in the Bermuda Triangle. Waterspouts happen when descending cloud combines with spray from the sea surface, to form a column of spinning water and air.

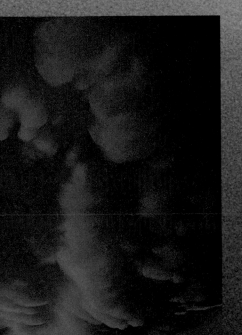

Well-known winds ❹

Berg South Africa	Hot, dry, downslope wind coming from the interior.
Bora Adriatic/Black Sea	Cold, dry, gusty north-east wind blows down from mountains at up to 200 km/h (120 mph).
Brickfielder Australia	Very hot summer wind carrying dust and sand.
Chinook USA	A warm, dry, turbulent wind blowing on the eastern side of the Rockies.
Levanter S. Alps and Mediterranean	Moist, strong, east wind that blows in summer and autumn.
Mistral South-east France	Dry, cold wind that blows down the Rhône valley into south-east France between autumn and spring. It can reach 100 km/h (60 mph).
Pampero South America	Cold wind blowing from the Andes across Argentina and Uruguay to the Atlantic.
Simoom Arabia, Egypt, Libya	Extremely hot, dry wind from the Sahara, up to 55°C (130°F). Can cause heatstroke.
Sirocco North Africa to southern Europe	Hot, dry, dusty wind blowing from Sahara across the Mediterranean. Known to cause lassitude. Humid in Europe.
Williwaw Alaska	Violent downslope wind blowing from snow-covered mountains to coast.

Going up ❺

Thermals occur when air coming into contact with sun-heated ground warms and rises, and colder air moves in to take its place.
◆ In most conditions, **updraught speeds** range from 0.5–1 m/sec (1½–3 ft/sec) to over 3 m/sec (10 ft/sec).
◆ Much more powerful updraughts can occur in **deep storm clouds**. In the 1950s the US Army and Weather Bureau recorded updraughts of 1000 m/min (3000 ft/min), and occasionally of 1500 m/min (5000 ft/min) in clouds that were 18000 m (60000 ft) high.
◆ **Gliders** use thermals to rise and need air current speeds of about 1 m/sec (2½–3 ft/sec) to gain sufficient lift.

Forest fires and fire storms ❸

Dry winds can combine with other local conditions such as drought and heat to spark a forest fire. Once alight, the fire can be driven at up to 15 km/h (9 mph) downwind. Sometimes a forest fire 'crowns': the fire spreads rapidly across upper foliage first, before the ground cover catches alight. The heat above draws the lower air upwards, creating fire-storm conditions of swirling winds and intense combustion.

WEIRD AND WONDERFUL ❼
Some **crop circles** – regular patterns found in wheat, rye or other grain fields – may be the result of a weather phenomenon called a plasma vortex in which a spinning mass of air has accumulated an electrical charge.

The water planet

Core facts ❶

◆ **Water**, more than anything else, makes the Earth different from the other planets.
◆ Seventy-one per cent of Earth's surface is covered by **oceans** and seas, and the oceans contain 97 per cent of the water on the planet.
◆ The remainder exists as **freshwater** in rivers, lakes, glaciers and underground.

◆ Freshwater also occurs in the atmosphere as **water vapour**.
◆ Excluding glaciers and ice caps, 94 per cent of liquid freshwater is **groundwater**.
◆ **A natural circulation** of water takes place between oceans, rivers, lakes, plants and the atmosphere.

The water cycle ❷

The circulation of water on Earth is **powered by the Sun**. Water in the oceans is heated by the Sun's rays, converted into water vapour and carried into the atmosphere by rising air currents. Ninety per cent of water in the atmosphere comes from the oceans, but moisture also evaporates from rivers, lakes and other surface water and from the respiration of plants and animals. Water vapour forms clouds, which deposit moisture as rain and snow back onto the surface. Water that collects on and under the ground flows back to sea again via streams and rivers.

MOISTURE LOSS Moisture is lost to the atmosphere from land and sea.

EVAPORATION The Sun heats the surface of the oceans, causing water to evaporate.

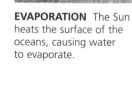

RETURN Water finds its way back to the sea via deltas or, as here, estuaries. Evaporation continues.

CLOUD FORMATION Water vapour in the air condenses into clouds.

Groundwater ❸

The ground is not as solid as it appears. When it rains, some water is held in the soil close to the surface; some percolates down into subsurface rock, where it fills up countless tiny cracks and spaces. The upper level of groundwater stored in rock is known as the water table. Springs occur where the surface cuts below the water table. Layers of rock which are permeable to water and store it easily are known as aquifers.

RIVERS These return water to the sea, and lose water directly to the atmosphere through evaporation.

965

Desalination

In desert areas salt can be removed from saline groundwater or seawater to make it fit to drink. Desalination is achieved either by filtering or distilling out the salt.

RAINFALL Moisture returns to land as rain or snow.

COLLECTION POINT Some rainfall seeps into the ground and emerges in springs and rivers.

COLLECTION POINT Lakes collect water from several sources, including rainfall, rivers and springs. They lose moisture through surface evaporation.

IRRIGATION CIRCLES A sophisticated and large-scale centre-pivot irrigation system maintains high maize yields in Colorado, USA.

World water supply

4

The total water content of Earth and the atmosphere is about 1.36 billion km³ (326 million cu miles). This works out at about 7.5 million litres (1.65 million gallons) of water for each person on Earth today. Human use of water is increasing each year. A washing-machine uses about 75 litres (16½ gallons) per cycle, a bath about 90 litres (20 gallons). A lawn sprinkler uses 1000 litres (220 gallons) per hour. The availability of water is variable and in some areas quality is deteriorating. Schemes to reverse these trends include water purification, reductions in industrial emission, desalination and sewage recycling.

THE DISTRIBUTION OF WATER

Oceans and open seas	97.2 per cent
Glaciers	2.15 per cent
Groundwater	0.62 per cent
Freshwater lakes	0.009 per cent
Salt lakes and inland seas	0.008 per cent
Soil moisture	0.005 per cent
Rivers	0.001 per cent
Atmosphere	0.001 per cent

Irrigation systems

5

In areas where rainfall is low or unreliable, irrigation can aid crop farming. **Primitive** irrigation in the form of dykes, channels and water lifting devices has been practised worldwide for centuries. **Modern** methods include centre-pivot systems and linear systems. Irrigation's downside is that it can result in the build-up of salts in the soil, rendering it useless.

Weather systems

Key facts ❶

◆The **atmosphere** over any continent is not homogenous, but consists of several **air masses**, some warm and some cold, that move around. The boundaries between two air masses of different temperatures or humidity are called fronts.
◆ Changes in weather from day to day are caused by the interactions between fronts.

◆ **Fronts** produce high and low-pressure areas at ground level, and these determine the type of weather at any time.
◆ **Cold fronts** travel faster than warm ones. An active cold front can travel at 50 km/h (30 mph).
◆ A combination of weather maps and satellite images provides a detailed picture of what the weather is doing.

Warm and cold fronts ❷

◆ **Warm fronts** bring warm air into cool areas. As a warm front moves in, it gradually rises over the cold air and cools. Clouds form, and rain falls over a wide area.
◆ **Cold fronts** bring cold air into warm areas. As warm air rises above cold air, a low-pressure area, or depression, forms at the ground. When a cold front overtakes a warm front inside a low-pressure area, an **occluded front** forms. Air begins to spiral in towards the centre, setting the system rotating. This produces unsettled weather and rain, and winds up to gale force.
◆ In Northern Hemisphere temperate zones, warm westerlies and cold polar winds set up an alternating pattern of cold and warm fronts moving across the North Atlantic.

THE COMING STORM The boundary between two air masses is clearly visible as this storm approaches.

★ 638
Squall lines

Sometimes a series of storms occurs in a line along an advancing front, which feeds the storms with moist air. In some cases exceptionally strong thunderstorms, known as supercell storms, form at the end of a squall line.

How the weather feels to us ❸

Weather forecasts distinguish between actual air temperature and how the weather feels to us. The cooling effect of the wind, known as **wind-chill factor**, is estimated by combining wind speed and air temperature. Relative **humidity** is a guide to the level of water vapour in the air. High humidity increases the apparent temperature.

How 'fronts' got their name ❹

The science of weather forecasting became established during World War I. Forecasters drew an analogy between the way warm and cold air masses continually jostle for position, and the interminable advances and retreats of armies on the **Western Front**. As a result, the boundaries between different air masses became known as fronts.

Weather maps

5

Data from satellites and weather balloons, land stations, ships and aircraft is used to draw up weather maps, which combine recent observations from several sources over a large area, and are known as synoptic charts.

Satellite images provide information on cloud formations and the locations of storm systems, although these are constantly moving and changing. On maps, this information is shown as areas of high (H) and low (L) pressure in conjunction with the positions of major fronts. **Isobars**, which are lines linking areas of equal air pressure, are shown on many maps. The air pressure is sometimes given.

Standard **symbols** for indicating fronts on a map are: triangles for a cold front; semicircles for a warm front; semicircles and triangles for an occluded front.

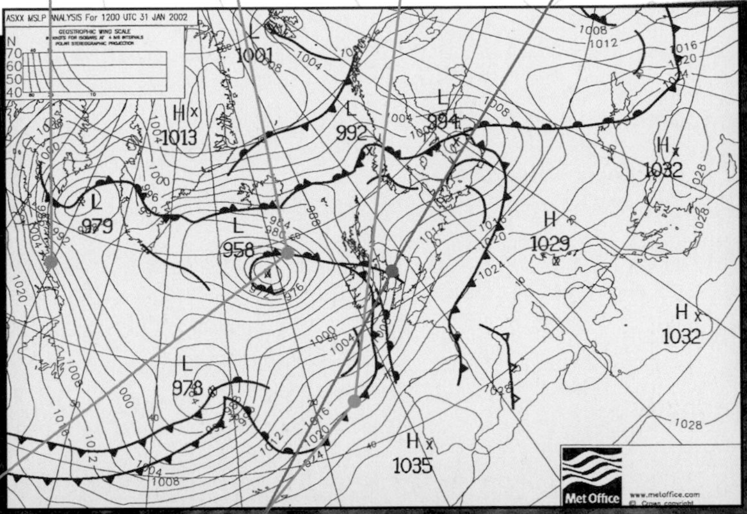

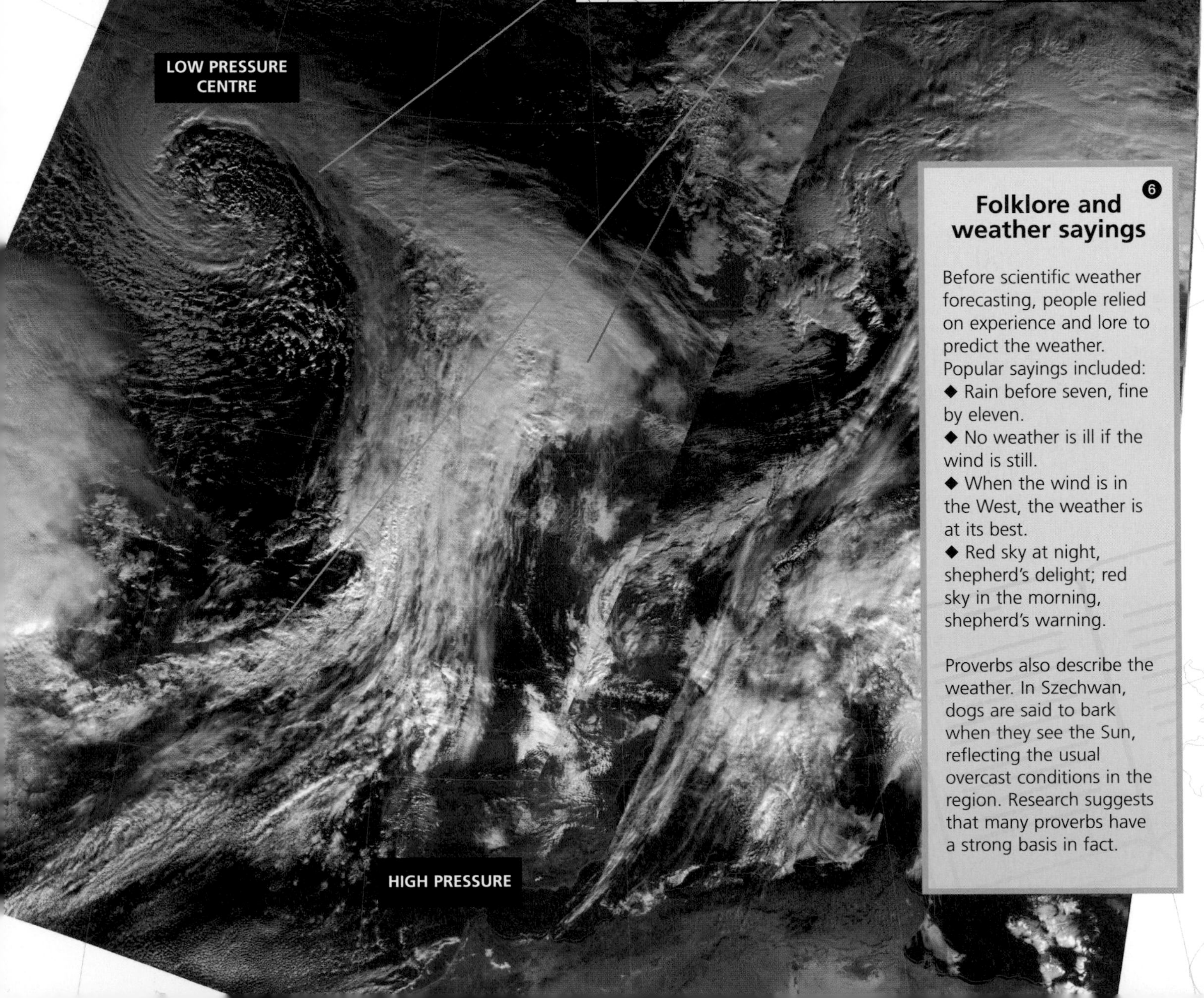

ISOBAR

OCCLUDED FRONT Indicates stormy weather.

COLD FRONT Leading edge of a cold air mass.

WARM FRONT Leading edge of a warm air mass.

LOW PRESSURE CENTRE

HIGH PRESSURE

Folklore and weather sayings

6

Before scientific weather forecasting, people relied on experience and lore to predict the weather. Popular sayings included:
◆ Rain before seven, fine by eleven.
◆ No weather is ill if the wind is still.
◆ When the wind is in the West, the weather is at its best.
◆ Red sky at night, shepherd's delight; red sky in the morning, shepherd's warning.

Proverbs also describe the weather. In Szechwan, dogs are said to bark when they see the Sun, reflecting the usual overcast conditions in the region. Research suggests that many proverbs have a strong basis in fact.

Storm systems

Core facts ❶

◆ The **thunderstorm** is one of the most common weather systems on Earth. Every day about 40 000 thunderstorms occur around the world; most occur over land in the tropics.

◆ Thunderstorms develop in **cumulonimbus** clouds, and are accompanied by heavy rain, hail, high winds and lightning, the electrical discharges that cause thunder.

◆ **Tropical storms** are known as hurricanes in America and the Caribbean. They are typhoons in the Pacific and China Sea, and tropical cyclones in the Indian Ocean and Australasia.

◆ Tropical storms **occur between** June and November in the Northern Hemisphere, and between November and May in the Southern Hemisphere.

Thunder and lightning ❷

Thunderstorms develop when warm, moisture-laden air rises rapidly, causing a cloud to expand upwards. Strong updraughts and downdraughts develop, causing a **separation of the electrical charges** inside the cloud: positive at the top, negative at the bottom. When the difference in electrical charge within a cloud or between the bottom of the cloud and Earth's surface is large enough, electricity is discharged.

Heat from a **lightning flash**, which can reach 22 000°C (40 000°F), causes the surrounding air molecules to expand rapidly, producing the sound waves we hear as thunder. Because light travels faster than sound, we see lightning before hearing thunder.

WEIRD AND WONDERFUL ❹

When lightning enters sand it can form thin tubes of glass, called **fulgurites**. Charles Darwin found many at the mouth of the River Plate. The tubes were 0.8-2.5 mm (^{1}/30-^{1}/10 in) thick, and one was nearly 2 m (6.6 ft) long.

20th-century storm records ❸

Cyclone 1970	The worst in terms of **number of deaths**: 300 000 were killed in Bangladesh.
Hurricane Gilbert 1988	The **largest hurricane**: reached 3500 km (2175 miles) in diameter.
Hurricane Mitch 1998	Overall, the most **destructive hurricane**: 9100 dead and a similar number missing; extensive flooding and mudslides in Honduras and Nicaragua.
Cyclone Tip 1979	The **largest tropical cyclone**: 2170 km (1350 miles) in diameter.

COUNTING THE COST A house in southern Florida devastated by Hurricane Andrew (1992), the most expensive storm in US history.

Naming hurricanes ❺

Hurricanes are given names to distinguish between weather systems that may exist in a region at the same time. Beginning in 1953, female names were routinely used. In 1975 Australia became the first country to use male names as well. The USA followed suit in 1978, using alternate male and female names. For Atlantic storms, names are allotted alphabetically, starting each year at A; they rarely get beyond M. In the north Pacific, they are named from A to Z.

⭐ 892

Types of lightning

Lightning discharges occur in three ways: within clouds, from cloud to air, or from a cloud to the ground. **Cloud-to-ground** lightning seeks the quickest route and may try several, which gives it its **forked** appearance. When a discharge takes place inside a dense cloud, we see the diffuse reflection on nearby clouds, known as **sheet** lightning. **Streak** lightning features one main bolt and many smaller shoots. **Bead**, **chain**, or **pearl-necklace** lightning consists of a string of highly luminous sections. **Rocket** lightning travels slowly (for lightning) and resembles a rocket-trail. **Ribbon** lightning follows a very bent path.

   St Elmo's fire is a bluish-green luminous discharge that appears on a ship's mast or aircraft's wings during a storm. It is named after the patron saint of sailors and is seen as a sign of good luck.

CLOUD TO CLOUD Unusually, a bolt of lightning streaks horizontally across the sky during an electric storm.

Tropical storms ❻

◆ A **mature hurricane** consists of bands of thunderstorms spiralling round a calm central area called the eye. An average hurricane can reach up to 970 km (600 miles) in diameter and contain hundreds of storms. They produce winds up to 250 km/h (150 mph), intense rain and ocean surges. They can last for weeks and travel thousands of miles.

◆ **Tropical storms** begin with intense convection of water over warm seas. As the storm begins to spin, set in motion by the Coriolis effect (page 132), the wind strengthens. When wind speed reaches 120 km/h (74 mph), the storm is classified as a hurricane (typhoon or tropical cyclone, see page 152).

◆ The **low-pressure area** at the storm centre draws up a mound of water, or swell, which runs ahead of the storm. This builds up as the storm approaches land and causes extensive flooding. Once a hurricane hits land it begins to weaken because its source of moisture is cut off.

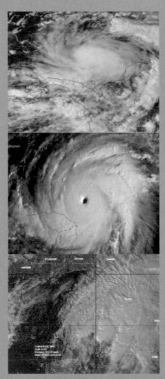

RISE AND FALL In three days Mitch grew from a tropical storm (top) to a hurricane (centre). The storm weakened when it hit the Florida coast.

Tropical storm paths ❼

Tropical storms develop over warm tropical seas. Hurricanes can only form in areas more than 10° latitude from the Equator, where the Coriolis effect (see page 132) begins to have influence and sets them rotating. Once rotating they move away from the Equator to the north or south.

   Some storms return towards the Equator. It is impossible for a hurricane to cross the Equator because the Coriolis effect loses its force and the storm stops rotating and decays.

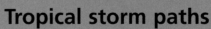

CURVING PATH Tropical storms travel west, then curve to the north in the Northern Hemisphere, south in the Southern Hemisphere.

Clouds

Core facts

◆ **Clouds form** when moisture in the atmosphere condenses into water droplets.
◆ **Fog** is, in effect, a type of ground-level cloud and also forms from condensation.
◆ **Clouds** are classified into two groups: rounded clouds, known as **cumulus**; and layered clouds, known as **stratus**. Clouds are also categorised by the height at which they occur above the ground. They reveal what is going on at different levels in the atmosphere and what type of weather may be on the way.
◆ A cloud's appearance also depends on factors such as temperature and wind speed, which can produce a range of variations.

Cloud formation ❷

Clouds are formed by rising air. There are three main causes:
◆ **Convection** A hot land surface heats the air, causing pockets of warm air to rise. This usually results in cumulus clouds.
◆ **Frontal cloud** Cold and warm fronts meet and the warm air is forced up. This usually results in stratus clouds.
◆ **Orographic lifting** Air is forced up the side of a mountain. This results in cumulus clouds.

As air rises, it cools. Cool air holds less water vapour than warm air, and it reaches a point when condensation takes place, forming water droplets or, at high altitudes, ice crystals.

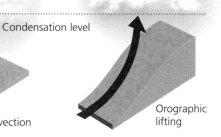

Condensation level

Convection

Orographic lifting

981

Fog

Officially, fog refers to conditions when visibility is no more than 1 km (0.6 miles). It forms from condensation close to the ground, either when warmer and cooler air currents meet, or when warm air flows over a cold surface. Sea fogs form where cold and warm currents meet and the air above them mixes. The thickness of fog depends on the quantity of water particles in the air.

MISTAKEN IDENTITY Smooth and layered, lenticular clouds form on the leeward side of mountains. Some have been mistaken for UFOs.

Cloud varieties and features ❸

Terms used to describe and distinguish between the many variations on basic cloud types include:

Anvil/incus A very tall, flat-topped cumulonimbus cloud – the characteristic shape of a thundercloud.
Castellanus The upper part of the cloud has projections resembling battlements. Applies to cirrus and all types of cumulus clouds.
Cloud Street Line of cumulus clouds formed parallel to wind direction.
Congestus A vertical cumulus cloud with bulging upper parts.
Lenticular Lens-shaped clouds that form over mountains. They remain relatively stationary.
Mackerel sky Cirrocumulus or small altocumulus clouds patterned like the scales of a mackerel.
Mother-of-pearl Wave clouds where light is diffracted into iridescent colours.
Noctilucent Rare, silvery luminous cloud seen on summer nights in high altitudes.
Undulatus Cloud that is undulating, or that has formed in waves.

Cirrus

CIRRUS High-level clouds composed of minute specks of ice. They form in feathery white wisps, and are sometimes known as mare's tails.

CIRROCUMULUS High-level cloud that forms in small, rounded tufts, sometimes in a 'mackerel sky' pattern.

Cirrocumulus

Cirrostratus

6000 m (19 500 ft)

CIRROSTRATUS High-level cloud that is stretched out in thin sheets.

Altocumulus

Common cloud types ❹

Clouds are categorised by their **shape** and by the **height** at which they form.

◆ Clouds that form in rounded heaps are called **cumulus**, and indicate unstable weather conditions.

◆ Clouds that form in flat layers are called **stratus,** and indicate stable conditions.

◆ Clouds that form above 6000 m (19 500 ft) are described by the prefix **cirro-** (or cirrus).

◆ Clouds that form 2000 m to 6000 m (6500 ft to 19 500 ft) are given the prefix **alto-** (or altus).

◆ For clouds that occur below 2000 m (6500 ft), **cumulus** and **stratus** are used on their own.

◆ **Nimbus** is used on its own, or in conjunction with other cloud names, and means 'rain-bearing'.

Altostratus

ALTOCUMULUS Mid-level cloud that is similar to cirrocumulus, with large tufts arranged in lines.

2000 m (6500 ft)

CUMULONIMBUS Have a flat base that occurs at around 1400 m (4500 ft). The top can extend to over 6 km (20 000 ft).

Stratocumulus

Stratus

STRATOCUMULUS Low-level sheets of cloud that produce a dull, grey sky.

CUMULUS Have a flat base and dome-shaped top. The base occurs at around 1400 m (4500 ft) and the top extends to about 1800 m (6000 ft).

Cumulus/Cumulonimbus

Sea level

Rain and snow

Core facts ❶

◆ **Rain**, **snow** and **hail** occur when water droplets or ice crystals in a cloud grow large and heavy enough to fall to Earth.
◆ The form in which moisture reaches the ground depends on the process of formation inside a cloud, and on the temperature between the cloud and the ground.

◆ **Frost** and **dew** form from the condensation of moist air when it comes in contact with a colder surface, such as the ground, plants or windows.
◆ In very cold conditions two types of freezing rain occur: ice pellets and glaze.
◆ No two **snowflakes** are identical.

Rain, snow and hail ❷

If the air temperature inside a cloud is above freezing, **water droplets** merge and grow until they are heavy enough to fall, reaching the ground as rain.

If **ice crystals** are present in a cloud, they bond together until large enough to fall. While falling, they either remain frozen, reaching the ground as snow, or melt, reaching the ground as rain.

Hailstones form in thunderclouds when several layers of ice build up on supercooled water droplets. Up to 25 layers have been recorded.

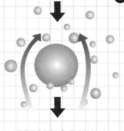

FORMING RAIN As a large water droplet falls through rising moist air inside a cloud, it captures smaller droplets and grows bigger.

	Average size	Largest known
Raindrops	0.5–5 mm (1/50–1/4 in)	Large ones break up
Snowflakes	10 mm (1/2 in) in diameter	38 cm (15 in) wide 20 cm (8 in) thick
Hailstones	Pea-sized	1 kg (2 1/4 lb)

Raining sand and frogs ❸

A number of accounts exist of creatures falling from the sky. In AD 77 the Roman historian Pliny described a shower of frogs. In 1859 a huge shower of fish fell in Glamorgan, Wales. In August 1918 in Hendon, Sunderland, it rained eels for 10 minutes. Such incidents were once regarded as portents or paranormal phenomena. It is now known that the strong updraughts, which can form in cumulonimbus clouds, tornadoes and waterspouts, occasionally pick up small creatures, or objects, carry them along and deposit them some distance away.

WEIRD AND WONDERFUL ❹

Attempts have been made to produce rain by **seeding clouds** with silver iodide crystals. These form ideal condensation nucleii, attracting water droplets until they are large enough to fall as rain.

POURING DOWN Heavy rain brings seasonal floods in Bangladesh.

Freezing rain

When supercooled raindrops fall through freezing air they form tiny **ice pellets** (known as sleet in the USA). The difference between ice pellets and hail is that ice pellets form in any type of cloud whereas hail only forms in thunderclouds.

When supercooled raindrops fall on cold solid surfaces such as roads and trees they spread out and freeze, covering surfaces with a sheet of clear or 'black' ice, called **glaze**. A heavy downpour in these conditions is called an **ice storm**.

ICE SCULPTURES An ice storm has smothered everything in its path in thick layers of ice.

WINDOW ART Hoar frost produces wonderful patterns on cold windows.

Frost ❺

Frost occurs when moist air comes into contact with a colder surface.
◆ **Air frost** When the air temperature is 0°C (32°F), or lower, as measured 1.2 m (4 ft) above the ground.
◆ **Fern frost** Feathery patterns that form on the inside of cold glass when dew freezes.
◆ **Hoar frost** Forms in damp air. The water vapour freezes directly onto a cold surface. The surface must be colder than the air and below freezing.
◆ **Ground frost** Hoar frost that forms on the ground or in the top layer of the soil and damages plants.
◆ **Rime** A crust of ice that forms when supercooled fog droplets freeze onto surfaces such as fences and trees. It can turn the whole landscape white.

FROZEN FOREST Hoar frost has coated the trunks of these fir trees during freezing conditions.

★ 753

Snowflake shapes

Snowflakes consist of **hexagonal** or six-pointed ice crystals. In very cold air, several crystals may clump together to form flakes that are columnar or needle-shaped.

American farmer W.A. Bentley photographed thousands of flakes through a microscope but never found two that were identical. But seven basic shapes have been identified:

| Plate | Column | Stellar | Spatial dendrite |

| Capped column | Irregular | Needle |

Atmospheric optical effects

Core facts ❶

◆ When sunlight strikes the atmosphere it is scattered, and sometimes split into its component colours, by water droplets, ice crystals and dust particles in the air, causing a variety of optical illusions.

◆ **Rainbows** and fogbows are formed by the bending (refraction) of light by water droplets.

◆ The **Northern** and **Southern Lights** (aurora borealis and aurora australis) that appear in the sky over the Poles are caused by charged solar particles reacting with the atmosphere.

◆ Intense heat can create **mirages** – optical illusions caused by refraction of light by heated layers of air close to the ground.

Rainbows and fogbows ❷

Raindrops bend (refract) light rays from the Sun, acting much like prisms by **splitting the light** into its seven component colours. Each colour emerges again at a slightly different angle depending on its wavelength, producing separate bands of colour. Red (the longest wavelength) appears on the outside of the rainbow, followed by orange, yellow, green, blue, indigo, and violet. The higher the Sun is in the sky, the flatter the rainbow. A **double rainbow** occurs when light is reflected within the raindrops.

When sunlight hits water droplets in fog an almost colourless **fogbow** results. The lack of colour is due to fine water droplets of fog not dispersing light as well as larger droplets.

Tie-breaker ❸

Q: Where would you see a 'green flash'?
A: On a rising or setting Sun. A green flash can occur at sunrise or sunset when some part of the Sun suddenly and briefly seems to **change colour** from red or orange to green or blue. This is an optical effect caused by the larger refraction of light at the blue/green end of the spectrum.

WEIRD AND WONDERFUL ❹

The world's **rainbow capital** is Honolulu in Hawaii. The mountains on the windward side of the islands cause the prevailing warm, moist trade winds to rise, creating clouds and almost daily rain, and forming lots of rainbows.

SUPER-SPECTRUM
A perfect rainbow
arches over the Mission
Mountains in Montana.

Sun effects ❺

Rainbows are just one optical effect caused
by sunlight passing through the air. Here
are some more.

Halo A circle of coloured light appears
around the Sun.

Parhelion (sundog, mock sun) A bright
spot appears on either side of the Sun.
Haloes and parhelia are caused by sunlight
refracting through ice crystals in the
atmosphere.

Corona A circle of light round the Sun,
caused by refraction through water
droplets.

Iridescence Light is diffracted into
patches of brilliant colour when it
passes through thin clouds.

SUNDOG A stunning parhelion
frames the legislature buildings
in Regina, Canada.

Sun pillars, subsuns White streaks and discs appear above the Sun, caused
by light being reflected by ice crystals rather than refracted through them.

Glory A highly magnified shadow of someone standing on a mountaintop is
cast on clouds below by a low Sun. Sometimes the shadow is surrounded by
bands of colour, as happens on the German mountain Brocken, where it is
called Brocken's Spectre.

Crepuscular rays Alternating dark and light rays are caused when sunlight is
split by an obstruction such as low cloud or a mountain.

⭐ **882**

Aurorae

These spectacular, shifting colours
seen at high latitudes are caused by
charged **solar-wind particles**
entering the upper atmosphere and
being deflected towards the Poles by
the Earth's magnetic field. As particles
spiral down the magnetic lines of
force they bombard gas molecules,
causing them to emit coloured light.

LIGHT DISPLAY Dancing aurorae lights
illuminate the northern night sky.

Mirages ❻

When sunlight passes through air layers of different temperatures and
densities, it bends (refracts). On very hot surfaces such as a desert floor or
tarmac road in summer, the light is refracted upwards. This produces a false
image (such as a sheet of water) just above ground level. Alternating layers of
cold and warm air can produce complex mirages that look like buildings
surrounded by water. They are called Fata Morganas, after the legendary
Morgan le Fay who was said to live in a castle under the sea. One famous Fata
Morgana appears in the Strait of Messina between Italy and Sicily.

SEEING THINGS Layers of air at different temperatures bend the light, creating an
image perceived by the viewer as floating above ground.

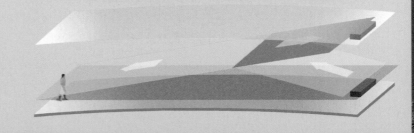

The changing atmosphere

Core facts ➊

◆ The **lower atmosphere** is in constant interaction with Earth's surface and biosphere.
◆ **Carbon**, an element essential to life, is continuously being recycled between the Earth and the atmosphere.
◆ A thin layer of **ozone** occurs in the stratosphere, where it absorbs much harmful ultraviolet radiation from the Sun. This layer is easily damaged by atmospheric pollution.
◆ **Fossil-fuel** burning is linked to problems of air pollution, acid rain and global warming.
◆ The **greenhouse effect** occurs when heat reflected from Earth's surface is prevented from escaping into space by gases in the atmosphere.

The carbon cycle ➋

Carbon dioxide is present in the atmosphere in relatively small quantities but its cycling between atmosphere and Earth is vital to life. It is taken up by plants during photosynthesis, converted into carbohydrate, transferred into animals by their ingestion of plant matter and released back into the atmosphere through plant and animal respiration. Fossil fuels are preserved plant and micro-organism remains and therefore contain carbon. When they are burnt, this is released as carbon dioxide. Volcanoes and the burning of large regions of tropical forests also produce carbon dioxide. The oceans absorb 25 per cent of all carbon dioxide released into the atmosphere.

ABSORBING AND RELEASING The oceans absorb carbon dioxide where some of it is used up by sea creatures to make calcium carbonate shells. These in turn, along with carbonate-rich sediments, can be turned into sedimentary rock. Carbon dioxide is released into the atmosphere by the weathering of these carbon-rich rocks such as chalk and limestone.

The greenhouse effect ➌

The main greenhouse gases are carbon dioxide (increased amounts created by burning fossil fuels and forests), nitrogen oxides (from vehicle exhausts), methane and water vapour. As the quantity of these gases increases, atmospheric temperature is expected to rise. Temperature has increased by 0.5°C (0.9°F) since 1905 and the United Nations predicts a further 1.5°C (2.7°F) increase by 2025, with a consequent sea level rise of 20 cm (8 in).

a Some of the Sun's heat that reaches the Earth is reflected back out again.

b As concentrations of greenhouse gases build up in the atmosphere, they re-reflect more solar heat back to Earth.

Ozone ❹

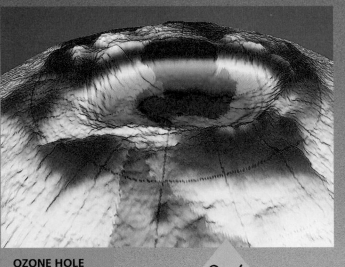

Most ozone is **produced** over the Tropics, where the Sun's radiation is strongest, and is distributed around the world by high-level winds. The ozone layer is **depleted** by increases in solar activity, by dust and gases released from volcanoes, and by long-lived pollutants such as CFCs, which destroy ozone.

The **hole** in the ozone layer over Antarctica first appeared in the 1980s. It forms because the cold air traps CFCs there in winter, then when the Sun returns, the combination of sunlight, CFCs and ice crystals produce an ozone-destroying mixture. **Depletion** over Artarctica was estimated at 50 per cent in the 1980s; ozone is depleting by 5 per cent every ten years over northern Europe.

OZONE HOLE
The blue patch in this satellite image shows the area of ozone depletion over Antarctica.

WEIRD AND WONDERFUL ❺

Mexico City's taxis add significantly to its lead pollution levels. An aggressive policy to reduce pollution is now visible in the number of **green-painted taxis** on the streets, indicating that they now run on cleaner, unleaded fuel.

Acid rain ❻

Waste gases such as sulphur dioxide and nitrogen oxides are released by industrial processes and by burning fossil fuels. They combine with water vapour in the atmosphere to form sulphuric and nitric acids. When precipitation occurs, the resulting rain or fog is highly acidic (pH 5.6 or lower) especially in Europe and the USA. Acid rain is damaging to plant life, freshwater creatures and building materials, and can travel large distances.

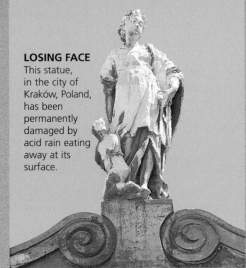

LOSING FACE
This statue, in the city of Kraków, Poland, has been permanently damaged by acid rain eating away at its surface.

Smog ⭐ 421

Smog (literally 'smoke + fog') has two distinct forms. **Sulphurous smog** is caused when smoke from fossil fuels, especially coal, combines with damp air. The Industrial Revolution caused severe sulphurous smog, notably the famous London 'peasoupers'. **Photochemical smog** results when hydrocarbon and nitrous oxide fumes, from vehicles and industrial processes, react with sunlight creating toxic ozone gas. Smog badly affects cities such as Los Angeles and Mexico City. Lanzhou, in China, can be so smog-bound it is invisible to satellites. Prolonged exposure to smog can cause respiratory problems, skin or eye irritation and lung disease.

SMOGGY MEXICO A pall of smog over Mexico City. Emissions from cars and industry react with sunlight to create high levels of ozone. This is exacerbated by the city's position in a bowl of mountains.

Altered Earth

Core facts ❶

◆ Earth's different environments consist of networks of interrelating **natural systems** which achieve a balance over time.
◆ Human activities, such as transport, farming, industry, dams and reservoirs, that alter the landscape are known as **footprints**. They can disrupt the balance of natural systems.

◆ **Waste materials**, such as effluent, rubbish and oil pollution, can never actually be disposed of, but enter the Earth's systems. It is not always possible to predict how natural processes will respond to their introduction.
◆ **Waste disposal** places one of the greatest strains on the natural environment.

Land and soil ❷

Soil is a vital element in the natural environment but it is very vulnerable to human activities. **Deforestation**, for instance, destroys root systems that bind the soil, so when trees are removed the soil is easily eroded away. Such areas are more susceptible to land and mudslides. **Farming** methods also degrade soil. Irrigation can cause a build-up of salt, making soil infertile, and large-scale farming practices can lead to soil erosion, as in 1930s America when drought created the **Dust Bowl**.

BACKGROUND IMAGE: Deforestation can have devastating effects. The cleared areas are particularly vulnerable to soil erosion.

Rivers and lakes ❹

Surface water is especially susceptible to pollution. **Runoff** from agricultural land feeds directly into rivers and lakes. Pesticides kill wildlife and fertilisers cause the overgrowth of aquatic plant life, using up oxygen and choking waterways. Sewage and industrial effluent are often discharged directly into waterways. The temperature of effluent is important as well as the content, as higher temperatures can kill aquatic life.

Dams and reservoirs ❸

Dam and reservoir construction is a common method of ensuring water supply, but there is an environmental cost. Construction can involve the relocation of settlements, as at the controversial Sardar Sarovar dam in India. It can cause destruction of wetlands, the cessation of natural flood-plain inundations with the loss of irrigation and natural fertilisation, the spread of diseases such as bilharzia and malaria, and the alteration of river and groundwater systems upstream.

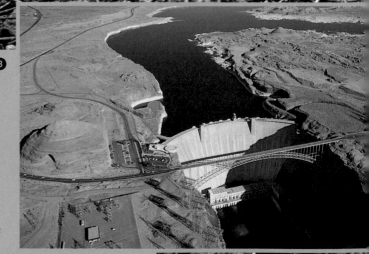

HOLDING BACK Glen Canyon dam on the Colorado River, Arizona.

A load of rubbish ❼

Rubbish disposal is one of the world's greatest environmental problems. The favoured method is **landfill**, but pollutants tend to leach from the rubbish into ground and surface water. The world's **largest landfill**, closed in 2001, is the 50-year-old, 890 hectare (2200 acre) Fresh Kills in Staten Island, NY, with four landfill mounds rising to 70m (225ft). In Tokyo, offshore islands have been created from rubbish landfills.

GARBAGE PILES Ninety per cent of the world's household waste is dumped in landfills like this one.

Tie-breaker ❺

Q: The *Amoco Cadiz* hit world headlines in March 1978 – what was it?
A: An oil tanker. *Amoco Cadiz* ran aground near Portshall, Brittany, on March 16, spilling 312.6 million litres (68.7 million gallons) of crude oil and polluting 386km (240 miles) of French coastline.

ACTIVE PROTEST Greenpeace combines direct action with scientific research.

Monitoring the Earth ❻

For some years now international initiatives and projects have been established to provide data that help environmentalists and governments try to rectify problems and prevent further damage. Landsat 7 is the latest of seven satellites launched between 1972 and 1999; its remote sensing scans provide vital images and data on the state of the Earth. The United Nations' **Earthwatch** programme, founded in 1972, circulates information and supports global environmental initiatives, acting as a catalyst for environmental monitoring.

★ 478

Oil spills

Oil spills, such as the *Exxon Valdez* spill in Alaska in 1989, endanger marine life by poisoning, by clogging feathers and gills, and by preventing sunlight from penetrating the water, reducing dissolved oxygen levels. Oil spills are cleared up by skimming, absorbing and chemical breakdown.

Scales and measurements

Core facts ❶

◆ The **main measurements** for weather are wind speed, temperature, air pressure, rainfall and humidity.
◆ Sir Francis **Beaufort**'s original wind force scale was based on observations of the effects of the wind on sailing ships.

◆ The **Kelvin** Scale is an absolute temperature scale based on absolute zero. Temperature intervals are the same as for degrees Celsius.
◆ Two different measures are used to describe an **earthquake**: the relative magnitude and the intensity of the tremors.

Weather ❷

Wind speed is usually measured using a cup **anemometer**, which consists of three or four cups mounted on a spindle. The stronger the wind, the faster the cups rotate. Simple models display the number of rotations. Electronic models provide a continuous wind-speed printout.

Beaufort Scale – wind strength and effects

Force	Wind speed	Description	Effect
0	Below 1 km/h (1 mph)	No motion	Smoke rises vertically
1	1–5 km/h (2–3 mph)	Light air	Smoke drifts slowly; leaves barely move
2	6–11 km/h (4–7 mph)	Light breeze	Drifting smoke indicates wind direction
3	12–19 km/h (8–12 mph)	Gentle breeze	Leaves rustle, wind felt on face
4	20–29 km/h (13–18 mph)	Moderate breeze	Leaves in constant motion; dust blows
5	30–38 km/h (19–24 mph	Fresh breeze	Small trees sway; paper blows away
6	39–51 km/h (25–31 mph)	Strong breeze	Large branches sway
7	52–61 km/h (32–38 mph)	Near gale	Whole trees sway
8	62–74 km/h (39–46 mph)	Gale	Tree twigs break; hard to walk
9	75–86 km/h (47–54 mph)	Strong gale	Branches break; roof tiles blow off
10	87–101 km/h (55–63 mph)	Whole gale	Small trees uprooted; roofs damaged
11	102–120 km/h (64–74 mph)	Storm	Widespread building damage
12	Above 120 km/h (74 mph)	Hurricane	Severe destruction

Saffir-Simpson Scale – hurricane strength

Category	Wind speed	Storm surge	Damage
1	120–152 km/h (74–95 mph)	1.1–1.6 m (4–5 ft)	Minimal
2	153–176 km/h (96–110 mph)	1.7–2.5 m (6–8 ft)	Moderate
3	177–208 km/h (111–130 mph)	2.6–3.7 m (9–12 ft)	Extensive
4	209–248 km/h (131–155 mph)	3.8–5.4 m (13–18 ft)	Extreme
5	More than 248 km/h (155 mph)	Above 5.4 m (18 ft)	Catastrophic

Fujita Scale – tornado strength

Scale number	Wind speed	Damage
F0	64–117 km/h (40–73 mph)	Light
F1	118–180 km/h (74–112 mph)	Moderate
F2	181–251 km/h (113–157 mph)	Considerable
F3	252–330 km/h (158–206 mph)	Severe
F4	331–417 km/h (207–260 mph)	Devastating
F5	More than 418 km/h (261 mph)	Incredible

Volcano explosivity

The **Volcano Explosivity Index** (VEI) describes the relative size of a volcanic eruption. It is based on factors such as the volume of material erupted and the height of the ash cloud, and consists of a scale from 0 to 8: 0 is non-explosive, 8 is megacolossal. The index is used to classify past eruptions and to indicate the potential scale of future eruptions. Vesuvius is on a VEI level 4 alert.

Asteroid hazards

The **Torino Impact Hazard Scale** indicates the hazard posed by newly discovered asteroids and comets. Each object is assigned a number from 0 to 10 and a colour code. An object assigned a 0 in the white zone is unlikely to collide with Earth or is so small it will break up in the atmosphere. A 10 in the red zone would mean certain collision and global climatic catastrophe.

❸

Temperature

There are three temperature scales: **Celsius** (centigrade), **Fahrenheit** and **Kelvin**. The Kelvin Scale is named after the physicist Lord Kelvin, who developed the absolute temperature scale. According to this, a theoretical absolute zero, or 0 K, is the baseline for measuring temperatures in the Universe.

Absolute zero (-273.15°C/-459.67°F) is the point at which all thermal energy disappears. It is the coldest possible temperature in the Universe.

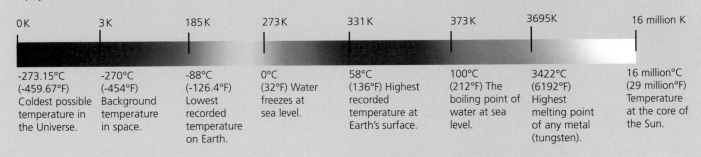

0 K	3 K	185 K	273 K	331 K	373 K	3695 K	16 million K
-273.15°C (-459.67°F) Coldest possible temperature in the Universe.	-270°C (-454°F) Background temperature in space.	-88°C (-126.4°F) Lowest recorded temperature on Earth.	0°C (32°F) Water freezes at sea level.	58°C (136°F) Highest recorded temperature at Earth's surface.	100°C (212°F) The boiling point of water at sea level.	3422°C (6192°F) Highest melting point of any metal (tungsten).	16 million°C (29 million°F) Temperature at the core of the Sun.

❹

Earthquakes

Richter scale

This estimates the **relative magnitude** of an earthquake based on the **largest seismic wave**. Large earthquakes produce waves thousands of times larger than weak ones. To accommodate this, Richter uses a logarithmic scale in which a tenfold increase in ground tremors corresponds to an increase of 1 on the scale. Also, each unit on the scale equates approximately to a 32-fold increase in the energy released. The Richter Scale has no upper limit.

Magnitude 1 — Detectable only by instruments.
Magnitude 2 — Barely detectable, even near the epicentre.
Magnitude 3 — Detectable near epicentre, but causes minimal damage.
Magnitude 4 — Detectable within 32 km (20 miles) of the epicentre.
Magnitude 5 — Shocks produce damage.
Magnitude 6 — Destructive in populated areas.
Magnitude 7 — Major earthquake. Serious damage.
Magnitude 8 — A great earthquake. Extensive destruction to communities near the epicentre.

Modified Mercalli scale

This scale grades earthquakes according to their **intensity**, which is assessed by the amount of damage they do.

I–II — Barely felt. Generally not recognised as an earthquake.
III–IV — Often felt. No damage, but suspended objects swing.
V–VI — Widely felt. Objects move, slight damage.
VII — Damage to poorly constructed buildings.
VIII — Damage to well-constructed buildings.
IX–X — Landslides. Wholesale destruction.
XI — Total damage. Visible ground movement.
XII — Total damage over large area. Objects thrown into the air.

HURRICANE FORCE
Hurricane Irene swept through Florida in 1999.

⭐ 512

Volcano-monitoring

Rising magma inside a volcano can produce cracks, swellings or subsidence on the volcano's flanks or dome. **Tiltmeters** are used to measure and monitor changes in the slope angle. This information, together with signs of earthquake activity and volcanic gases, are used to provide eruption warnings.

The page number in brackets indicates where you will find the questions for each quiz.

Quiz 0 (page 8)

1 Earth
2 Slate
3 Black Forest gâteau
4 Vermont
5 Blue
6 The Arctic
7 Outer space
8 Yellow
9 Greenpeace
10 Lead white

Quiz 1 (page 8)

11 Tsunami
12 Troposphere
13 Gulf Stream
14 Sunlit, twilight, abyssal, hadal
15 USA and Canada
16 Fogbow
17 Glaciers
18 South to north
19 Lakes
20 It has none

Quiz 2 (page 8)

21 Hawaii
22 The Balearic Islands
23 *Swiss Family Robinson*
24 The Seine
25 Christmas Island
26 Greece
27 HMS *Bounty*
28 Almost an island
29 Ellis Island
30 Atlantis

Quiz 3 (page 8)

31 Isle of Man
32 Mice
33 Silver
34 The dog was called Snowy
35 Huskies
36 Cockerel
37 The desert
38 In coal mines
39 Canary
40 A fish

Quiz 4 (page 9)

41 Antarctica
42 Asia
43 North America
44 Africa
45 Antarctica
46 Oceania (Australia)
47 South America
48 Asia
49 Asia
50 Oceania

Quiz 5 (page 9)

51 Earth
52 A comet
53 Volcano eruption and tidal wave
54 *Mars Attacks!*
55 Tornadoes
56 Gold
57 A waterfall
58 Albert Einstein
59 *Galaxy Quest*
60 The Rockies

Quiz 6 (page 9)

61 D
62 B
63 A
64 A
65 A
66 A
67 A
68 A
69 B
70 C

Quiz 7 (page 10)

71 Sunglasses
72 Sundial
73 Sunflower
74 Sunspots
75 Sun roof
76 Sunbow
77 Sundance
78 Sun King
79 Sunstone
80 Sunblock

Quiz 8 (page 10)

81 False
82 True
83 False
84 True
85 False
86 False
87 True
88 True
89 False
90 True

Quiz 9 (page10)

91 Ultraviolet
92 Greenwich Mean Time
93 CFCs
94 Acidity
95 Nitrogen oxides
96 Self-contained underwater breathing apparatus
97 Very hot
98 Extra-vehicular activity
99 Lunar module
100 Sonar

Quiz 10 (page 10)

101 San Andreas
102 St Francis
103 Epicentre
104 Aftershock
105 It sank
106 Four minutes
107 Poseidon
108 14.5 m (47$^{1}/_{2}$ ft)
109 Colossus of Rhodes
110 Charles Darwin

Quiz 11 (page 11)

111 Tide
112 Africa
113 Chuck Norris
114 *On Her Majesty's Secret Service*
115 A coalminer
116 Marlon Brando
117 Universe
118 Oil
119 Tides
120 Sahara

Quiz 12 (page 11)

121 Neptune
122 Sun orbited Earth
123 In a great flood
124 Air, water, earth, fire
125 The sun god
126 Moonstone
127 The Moon
128 A lake
129 Himalayas
130 Bottom of the sea

Quiz 13 (page 11)

131 B
132 D
133 B
134 A
135 D
136 B
137 D
138 A
139 D
140 D

Quiz 14 (page 12)

141 Buzz Lightyear
142 Nuclear
143 Vincent van Gogh
144 Solar eclipse
145 Pot of gold
146 Thor
147 Natural gas
148 It's the same mountain
149 Towards each other
150 DDT

Quiz 15 (page 12)

151 Andes
152 Venice
153 Reservoir
154 Mirage
155 Gobi
156 Monsoon
157 Meteorite
158 Dune
159 Pyrenees
160 Avalanche

Quiz 16 (page 12)

161 Windy city
162 Backbone of Italy
163 Dirty snowball
164 Big Bang
165 Emerald Isle
166 Craters of the Moon
167 St Elmo's fire
168 Death Valley
169 Poor man's space probe
170 Great Bear

Quiz 17 (page 12)

171 Tableland
172 Precambrian
173 Halo
174 A meander
175 Tundra
176 Wadi
177 Rainshadow
178 Abrasion
179 Laurasia and Gondwana
180 Hamada

Quiz 18 (page 13)

181 Guilin Hills
182 Cappadocia
183 Devil's Marbles
184 Devil's Tower
185 Uluru
186 Giant's Causeway
187 Delicate Arch
188 Wave Rock
189 Grand Canyon
190 The Mittens

Quiz 19 (page 14)

191 Waterfalls
192 Water Carrier
193 Watergate
194 Waterspout
195 *Waterworld*
196 Waterline
197 Watershed
198 Water butt
199 Water cycle
200 *Waterfront*

Quiz 20 (page 14)

201 Mars
202 *The Tempest*
203 *Gone with the Wind*
204 Rain fell upon the Earth
205 *Little House on the Prairie*
206 Snow
207 In a tornado
208 India
209 *Sea*
210 Diamond

Quiz 21 (page 14)

211 Baseball
212 He skied
213 Silver
214 Golf
215 Venus Williams
216 Slate
217 Nimbus
218 Surfing, surfboarding
219 Hurricane Higgins
220 *Avalanche*

Quiz 22 (page 14)

221 Wind gauge
222 Wind farm
223 Wind pack
224 Wind chill
225 Windward Islands
226 Windsurfing
227 Windsock
228 Wind scale
229 Wind gap
230 Wind stress

Quiz 23 (page 15)

231 Northern
232 *The Lost World*
233 Continental shelf
234 Landslide
235 A type of wind
236 A picnic
237 1944
238 Niagara Falls
239 Lake Geneva
240 Australia

Quiz 24 (page 15)

241 False
242 True
243 True
244 True
245 False
246 True
247 True
248 False
249 True
250 True

Quiz 25 (page 15)

251 C
252 C
253 D
254 A
255 B
256 C
257 D
258 A
259 D
260 B

Quiz 26 (page 16)

261 Hot
262 Cold
263 Cold
264 Cold
265 Hot
266 Hot
267 Hot
268 Cold
269 Cold
270 Hot

Quiz 27 (page 16)

271 Pole star
272 Lone Star State
273 Telstar
274 Shooting star
275 Flare star
276 Morning Star
277 Binary star
278 Dog Star
279 Lodestar
280 Dark star

Quiz 28 (page 16)

281 Italy
282 The Sun melted his wax wings
283 Hail
284 Mineral water
285 Basalt
286 Charles Blondin
287 An underwater hydrothermal vent
288 Apart
289 It's a lake of pitch or asphalt
290 Hydroelectricity

Quiz 29 (page 16)

291 Ol' Man River – the Mississippi
292 A star
293 The Comets
294 Summertime
295 Jupiter
296 Cloud
297 *South Pacific*
298 Rain
299 Moon River
300 A foggy day

Quiz 30 (page 17)

301 Constellations
302 Cape
303 Sea
304 Wind
305 Space probes
306 Dune
307 Earthquake measurement
308 Plain
309 Moons of Jupiter
310 Bay

Quiz 31 (page 17)

311 Atlas
312 Yogi Bear
313 Oort Cloud
314 Etna
315 Stimpy
316 Suez
317 Pluto
318 Karakoram
319 Capricorn
320 Richter

Quiz 32 (page 17)

321 Cloud
322 April
323 Planet
324 The desert
325 Richard Nixon
326 Paul Simon
327 South Pole
328 Lightning
329 Hail
330 Apollo 13

Quiz 33 (page 17)

331 Gibraltar
332 Metamorphic
333 Igneous
334 Sedimentary
335 Alcatraz
336 Chalk
337 Lava
338 Greenish-brown
339 Flint
340 On the rocks

Quiz 34 (page 18)

341 Water
342 Land
343 Water
344 Land
345 Land
346 Water
347 Land
348 Water
349 Land
350 Land

Quiz 35 (page 18)

351 *Apollo 13*
352 *Death on the Nile*
353 *Stagecoach*
354 *Diamonds are Forever*
355 *Close Encounters of the Third Kind*
356 *Dante's Peak*
357 *Star Wars: Phantom Menace*
358 *The China Syndrome*
359 *Contact*
360 *Deliverance*

Quiz 36 (page 18)

361 Clay
362 Little Rock
363 Marble
364 Jewels
365 Pearl
366 Pumice stone
367 Limestone
368 Granite
369 Pebble
370 Climbing mountains

Quiz 37 (page 18)

371 Red sky
372 Black
373 Blue Ridge Mountains
374 Stars
375 Black hole
376 The Yellow River, or Huang He
377 The Blue Mountains
378 Edward White
379 Rubicon
380 Black ice

Quiz 38 (page 19)

381 Reykjavik
382 El Capitán
383 Kilimanjaro
384 Victoria Falls
385 Cretaceous
386 Lithosphere
387 Mesa Verde
388 Okenofee
389 Pangaea
390 Thermosphere

Quiz 39 (page 19)

391 Carbon monoxide
392 South America, Africa and Asia
393 Light year
394 Less
395 Finland
396 Solar wind
397 1906 earthquake
398 Atmospheric pressure
399 Sweden and Denmark
400 Distance

Quiz 40 (page 19)

401 A
402 C
403 C
404 D
405 B
406 B
407 D
408 A
409 C
410 C

Quiz 41 (page 20)

411 Tiber
412 Amazon
413 Shannon
414 Danube
415 Kwai
416 The East or the Hudson
417 *Death on the Nile*
418 Thames
419 Mississippi
420 Suwannee/Swanee

Quiz 42 (page 20)

421 Smog
422 Mount Ararat
423 *A Passage to India*
424 Stephen Hawking
425 Halley's Comet
426 William Shakespeare
427 *Jurassic Park*
428 20 000 leagues
429 *Total Eclipse*
430 Robert Louis Stevenson

Quiz 43 (page 20)

431 February 29
432 Three Mile Island
433 80 days
434 Seven
435 360
436 70 per cent
437 Forty-niners
438 12
439 Seven
440 *Deep Space Nine*

Quiz 44 (page 20)

441 An avalanche
442 California
443 Mount Fuji
444 Meteorites reach the Earth's surface
445 Morocco, Algeria and Tunisia
446 Table Mountain
447 Boreas
448 A rock-weathering process
449 On the ocean floor
450 Depression

Quiz 45 (page 21)

451 Moon's surface
452 Supernova
453 Comet
454 Spiral galaxy
455 Mars' surface
456 Betelgeuse
457 Shooting star
458 Nebula
459 Neptune
460 Asteroid

Quiz 46 (page 22)

461 Beaches
462 Galaxy
463 Cloud
464 Space stations
465 Frost
466 Year
467 Pesticide
468 Wave
469 Space telescopes
470 Hole

Quiz 47 (page 22)

471 A mirage
472 Red, orange, yellow, green, blue, indigo, violet
473 Neap
474 Uluru
475 Typhoon
476 Halley's Comet
477 Turkey
478 The *Exxon Valdez*
479 Lake
480 Expand

Quiz 48 (page 22)

481 The Americas were named after him
482 Mars
483 Roald Amundsen
484 Pacific, Atlantic
485 Botany Bay
486 Uranus
487 David Livingstone
488 Saturn
489 Trinidad
490 11 km (6¾ miles)

Quiz 49 (page 22)

491 Phoenix
492 Hercules
493 Gemini
494 Taurus
495 Aquila
496 Orion
497 Hydra
498 Delphinus
499 Centaur
500 Cygnus

Quiz 50 (page 23)

501 Snow
502 Every rainbow
503 Hurricane
504 Marilyn Monroe
505 Wand'rin' star
506 On the Moon
507 Diamonds
508 In a cave
509 Raindrops
510 Island Records

Quiz 51 (page 23)

511 Earthquakes
512 Tiltmeter
513 Liquid
514 Barometer
515 Height above sea level
516 Salt in water
517 High temperatures
518 Humidity
519 Wind speed
520 Earth slippage

Quiz 52 (page 23)

521 C
522 A
523 B
524 C
525 B
526 B
527 A
528 C
529 C
530 B

Quiz 53 (page 24)

531 Ice age
532 Iceland
533 Ice shelf
534 Iceberg
535 Icefall
536 Ice blink
537 Ice-breaker
538 Ice sheet
539 Ice jam
540 Ice needles

Quiz 54 (page 24)

541 On the Moon
542 Deep sea
543 North Pole
544 Hovercraft
545 Balloon
546 Venice
547 Aeroplane
548 An iceberg
549 Polar ice cap
550 West wind

Quiz 55 (page 24)

551 Nile
552 Mediterranean
553 Hydrogen
554 China
555 Debris
556 Tyrannosaurus
557 Olympia
558 Nitrogen
559 New Orleans
560 Kilimanjaro

Quiz 56 (page 24)

561 Copper
562 Cullinan
563 The phases of the Moon
564 Venus
565 Gold
566 Kelvin
567 Hallmark
568 Water
569 Emerald
570 Crux, or the Southern Cross

Quiz 57 (page 25)

571 Table Mountain
572 Mount Everest
573 Mount Rushmore
574 Mount St Helens
575 Mount Fuji
576 Mount Kilimanjaro
577 The Matterhorn
578 Popocatépetl
579 Mount Erebus
580 Heimaey

Quiz 58 (page 26)

581 Mir
582 Lightning
583 Mediterranean
584 Fjord
585 Maelstrom
586 Delta
587 Sirocco
588 Igloo
589 Nebula
590 Tundra

Quiz 59 (page 26)

591 Landslide
592 Land breeze
593 Landlocked
594 Landmass
595 Land yacht
596 Landmark
597 Land bridge
598 Landfill
599 Land's End
600 Landsat

Quiz 60 (page 26)

601 True
602 False
603 True
604 True
605 True
606 False
607 True
608 False
609 False
610 True

Quiz 61 (page 26)

611 Platinum
612 Carat
613 Mercury
614 Fool's gold
615 Tin and lead
616 Tungsten
617 Aluminium
618 Nickel
619 Panning for gold
620 Fort Knox

Quiz 62 (page 27)

621 *Earthquake*
622 *Scott of the Antarctic*
623 *Titanic*
624 *The River Wild*
625 *The Dish*
626 *Planet of the Apes*
627 *Lawrence of Arabia*
628 *Southern Comfort*
629 *The World Is Not Enough*
630 *Jurassic Park*

Quiz 63 (page 27)

631 Red, black, yellow, green and blue
632 Greenhouse effect
633 Blue
634 White cap
635 Great Red Spot
636 Oil
637 Greenland
638 White squall
639 Purple Heart
640 Infrared

Quiz 64 (page 27)

641 D
642 B
643 C
644 A
645 C
646 B
647 C
648 B
649 D
650 D

Quiz 65 (page 28)

651 Autumn
652 Albert Einstein
653 Cloud
654 Neil Armstrong
655 Earth
656 Volcano
657 Mineral
658 June
659 Harvest Moon
660 Sting

Quiz 66 (page 28)

661 Mount Rushmore
662 Monte Carlo
663 The Eiger
664 Mount Olympus
665 Alps
666 *Fantasia*
667 Montana
668 Chris O'Donnell
669 Wyoming
670 Mount Sinai

Quiz 67 (page 28)

671 Oil
672 Farming
673 Volcanoes
674 Dinosaurs
675 Landslides
676 National parks
677 Wetlands
678 Orbits
679 Galaxy
680 Cave systems

Quiz 68 (page 28)

681 No
682 Inside the Arctic or Antarctic circle
683 Monsoon
684 Crust
685 Amazon rain forest
686 Snow
687 Dyke
688 A small avalanche
689 A volcanic eruption
690 Pulsars

Quiz 69 (page 29)

691 Tibet
692 Big crunch
693 Poor man's weatherglass
694 Pegasus
695 Jupiter, Saturn, Uranus, Neptune
696 London
697 The Land of the Rising Sun
698 Antarctica
699 A wind
700 Slash-and-burn

Quiz 70 (page 29)

701 Weather balloon
702 Temperate
703 Atmospheric or barometric pressure
704 An occluded front
705 Thermals
706 Front
707 Fine
708 Synoptic chart
709 Wind-chill factor
710 A cold front

Quiz 71 (page 29)

711 D
712 A
713 A
714 D
715 C
716 A
717 C
718 B
719 D
720 C

Quiz 72 (page 30)

721 Earthquake
722 Earthshine
723 Earth pillar
724 Earthwatch
725 Earthfill
726 Earth flow
727 Earth sciences
728 Earthshaker
729 Earthrise
730 Earthwork

Quiz 73 (page 30)

731 The core
732 Mississippi
733 Sinking of the *Titanic*
734 Sir Isaac Newton
735 Three dams
736 Kalahari
737 The Sun
738 Cave is horizontal, pothole is vertical
739 Acid fog
740 Supernova

Quiz 74 (page 30)

741 Falkland Islands
742 It can fly
743 Avalon
744 The poet John Donne
745 Napoleon
746 Devil's Island
747 Robinsoe Crusoe
748 Sicily
749 The Canaries
750 Easter Island

Quiz 75 (page 30)

751 Spock
752 40 days and nights
753 Six
754 Eight minutes
755 One
756 24
757 Roaring Forties
758 Hannibal
759 Antonio Vivaldi
760 67.2 m (220 ft)

Quiz 76 (page 31)

761 Canada
762 Hawaii
763 England
764 Mexico
765 Chile
766 Alaska
767 Italy
768 Australia
769 Somalia
770 India

Quiz 77 (page 32)

771 The Great Lakes
772 France and Switzerland
773 *On Golden Pond*
774 A volcano
775 Lake Titicaca
776 Lake Victoria
777 Lake Poets
778 By Great Salt Lake
779 Loch Ness
780 Aswan High Dam

Quiz 78 (page 32)

781 In the wind
782 On a tree by a river
783 *Summer Holiday*
784 Prairie
785 Lightning ('Greased Lightning')
786 Rain
787 *Thunderball*
788 'Great balls of fire!'
789 Hurricanes
790 Dark side

Quiz 79 (page 32)

791 Continents
792 Power or energy
793 Star
794 Forests
795 Layer
796 Observatories
797 Rock
798 Climates
799 Landslides
800 Belt

Quiz 80 (page 32)

801 North and South
802 South
803 North Sea
804 *North by Northwest*
805 North face
806 North-east
807 South America
808 From north-east to south-west
809 Magnetic north
810 In the west

Quiz 81 (page 33)

811 A volcano
812 A jewel with a panther-shaped flaw
813 *Godzilla*
814 A river
815 *Full Metal Jacket*
816 An asteroid
817 *The Tempest*
818 *Fahrenheit 451*
819 *Sunset Boulevard*
820 *Blue Thunder*

Quiz 82 (page 33)

821 Los Angeles
822 Fire/metalworking
823 Caldera
824 *Journey to the Centre of the Earth*
825 *The Little Prince*
826 Quiet period between eruptions
827 Edinburgh
828 Pompeii, Herculaneum
829 *Joe Versus the Volcano*
830 Mars

Quiz 83 (page 33)

831 D
832 C
833 B
834 B
835 C
836 A
837 B
838 B
839 A
840 B

Quiz 84 (page 34)

841 White
842 Red
843 White
844 Blue
845 Red
846 Red
847 Blue
848 White
849 Blue
850 Red

Quiz 85 (page 34)

851 Prehistoric wall paintings
852 Open sesame
853 Upwards
854 Batman
855 555km (345 miles)
856 Ice
857 It is blind
858 Dead Sea Scrolls
859 Cheddar
860 Speleology

Quiz 86 (page 34)

861 San Francisco
862 Niagara Falls
863 Thames
864 Seine
865 Florence
866 Japan's Inland Sea
867 Hudson
868 London
869 Mississippi
870 Rhine

Quiz 87 (page 34)

871 True
872 True
873 True
874 True
875 False
876 False
877 True
878 False
879 False
880 True

Quiz 88 (page 35)

881 Crepuscular rays
882 Aurora
883 Lenticular cloud
884 Tornado
885 Sun halo
886 Cumulus cloud
887 Glory
888 Cumulonimbus cloud
889 Sun pillar
890 Cirrus cloud

Quiz 89 (page 36)

891 Comets
892 Lightning
893 Ice
894 Mountain ranges
895 Canals
896 Astronomers
897 Fossils
898 Seas
899 Deserts
900 Reefs

Quiz 90 (page 36)

901 Frogs
902 600km (360 miles)
903 Subsoil
904 Summer
905 Humidity
906 Asbestos
907 Ozone layer
908 Kryptonite
909 Flood plain
910 Beneath a volcano

Quiz 91 (page 36)

911 Nile
912 Rhône
913 Mekong
914 Mississippi
915 Chang Jiang
916 Tiber
917 Neva
918 Orinoco
919 Volga
920 Yodo

Quiz 92 (page 36)

921 Thunder
922 *Seasons*
923 A desert island
924 Summer
925 Gold
926 A planetarium
927 Cloud Cuckoo Land
928 To make it safe from earthquakes
929 Herculaneum
930 Lightning

Quiz 93 (page 37)

931 Algae
932 Black frost
933 Brown
934 Green
935 Orange
936 Copper
937 Yellow Sea
938 Water
939 Blue whale
940 Sulphur

Quiz 94 (page 37)

941 Radioactivity
942 Longitude
943 Swedish
944 Submarine
945 Jacques Cousteau
946 Telescope
947 Suntan cream
948 Gunpowder
949 Ernest Rutherford
950 Lightning conductor

Quiz 95 (page 37)

951 A
952 D
953 A
954 C
955 A
956 B
957 D
958 A
959 C
960 D

Quiz 96 (page 38)

961 Saltwater
962 Brine
963 Lot's
964 3 per cent
965 Desalination
966 Salary
967 Sodium chloride
968 Rock salt
969 Warm
970 Raise

Quiz 97 (page 38)

971 Pacific
972 Mercury
973 Hurricanes
974 Mont Blanc
975 Aurora borealis
976 Greenland
977 Smoke and fog
978 Cumulus
979 The Doldrums
980 El Niño

Quiz 98 (page 38)

981 Fog
982 Frost
983 Diamonds
984 Rainbow
985 'Colors of the Wind'
986 'Moonlight Sonata'
987 Mountain
988 Subterranean
989 The white cliffs of Dover
990 *The Rite of Spring*

Quiz 99 (page 38)

991 −89.2°C
992 3000m
993 365.24219
994 1622
995 9.5
996 807m
997 200 billion
998 98 per cent
999 2172150km/h
1000 4.6 billion years

Question sheet

Quiz Number   **Quiz Title**

Questions	Answers
1	1
2	2
3	3
4	4
5	5
6	6
7	7
8	8
9	9
10	10

Answer sheet

Name

Quiz Number	Quiz Title

Answers

1

2

3

4

5

6

7

8

9

10

Total score

Abbreviations:
T = top; M = middle; B = bottom;
L = left; R = right

Front cover: Delicate Arch, Arches National Park, USA – ImageState.
Back cover: Sun prominences – Michael Holford.

1 Science Photo Library/National Optical Astronomy Observatories. **2–3** DRK Photo/Stephen J. Krasemann. **4–5** Science Photo Library/Soames Summerhays. **13** top to bottom: DRK Photo/Pete Oxford, 1; Roy Williams, 2; Digital Vision, 3; DRK Photo/Tom & Pat Leeson, 4; DRK Photo/Michael Fogden, 5; Auscape/Tom Till, 6; DRK Photo/Jeremy Woodhouse, 7; Auscape/Jean–Paul Ferrero, 8; Science Photo Library/ William Ervin, 9; Science Photo Library/ Alex Bartel, 10. **21** top to bottom/left to right: Science Photo Library/Richard J. Wainscoat, Peter Arnold Inc, 1; Science Photo Library/Pekka Parviainen, 2; Science Photo Library/National Optical Astronomy Observatories, 3; Science Photo Library/Space Telescope Science Institute/NASA, 4; NASA, 5; Science Photo Library/US Geological Survey, 6; Science Photo Library/ Voyager 2/NASA, 7; Science Photo Library/Celestial Image Company, 8; Science Photo Library/John Chumack, 9; Science Photo Library/NASA, 10. **25** top to bottom/left to right: Gerald Cubitt, 1; DRK Photo/David Woodfall, 2; DRK Photo/Wayne Lankinen, 3; Science Photo Library/Prof. Stewart Lowther, 4; Trip/A. Tovy, 5; Auscape/Ferrero-Labat, 6; John Cleare/Mountain Camera, 7; Science Photo Library/Wesley Bocxe, 8; Auscape/Jean–Paul Ferrero, 9; Science Photo Library/KRAFFT/HOA–QUI, 10. **31** Bradbury and Williams. **35** top to bottom/left to right: Digital Vision, 1,4; DRK Photo/Johnny Johnson, 2; Science Photo Library/Magrath Photography, 3; DRK Photo/Stephen G. Maka, 5; DRK Photo/Tom Bean, 6,7,8; DRK Photo/S. Nielsen, 9; DRK Photo/Don & Pat Valent, 10. **40** Matthew White, BL. **40–41** Julian Baker. **41** Science Photo Library/NASA, TL. **42** Science Photo Library/National Optical Astronomy Observatories, TR, ML; Science Photo Library/Celestial Image Co, MM. **43** Julian Baker, TR; Science Photo Library/Frank Zullo, MR; Science Photo Library/Mount Stromlo and Siding Spring Observatories, BL. **44** Bradbury and Williams. **44-45** Auscape/Tim Acker. **45** Anglo–Australian Observatory/photograph by David Malin, TL. **46** Julian Baker. **46–47** Laurence Bradbury. **47** Science Photo Library/ESA, TL; Science Photo Library/National Optical Astronomy Observatories, TR; Science Photo Library/NASA, ML; Michael Holford, BR. **48** Bradbury and Williams. **49** Science Photo Library/NASA/JPL, TR; NASA/JPL, MM. **50** Science Photo Library/US Geological Survey, TR; NASA/JPL. TL, MR; Science Photo Library/NASA, BR. **51** Science Photo Library/NASA, TM; NASA/JPL, ML, MR; Science Photo Library/NASA/Space Telescope Science Institute, BL, BR. **52** Galaxy Picture Library, MM; Science Photo Library/NASA, BM. **52–53** DRK Photo/Tom Bean. **53** Bradbury and Williams, ML; NASA, MR. **54** Matthew White, MM; Bradbury and Williams, B. **54–55** Science Photo Library/Tony and Daphne Hallas. **56** Julian Baker. **56–57** Science Photo Library/A. Behrend/Eurelios. **57** Science Photo Library/David Parker, TL; Science Photo Library/Detlev van Ravenswaay, TR. **58** Bradbury and Williams/Mountain High, BM. **58–59** Julian Baker. **59** Matthew White, TL, BR; Matthew White/Mountain Camera, TR. **60** Science Photo Library/Bernhard Edmaier, MR; Matthew White. **61** Bradbury and Williams, TL; NASA/GSFC/IMAGE, TL; Matthew White, MR; Bradbury and Williams/Mountain Camera, B. **62** Bradbury and Williams, MM; DRK Photo/Stephen J. Krasemann, BR. **62–63** Science Photo Library/Adam Jones. **63** Auscape/Jaime Plaza Van Roon, TL; DRK Photo/J. Wengle, BR. **64** DRK Photo/Stephen J. Krasemann, MM; DRK Photo/Jeremy Woodhouse, B. **64–65** Science Photo Library/Dr Juerg Alean. **65** DRK Photo/Lewis Kemper, TL; Matthew White, BL. **66** Auscape/Jean-Paul Ferrero. **66–67** DRK Photo/Tom Bean. **67** DRK Photo/Michael Collier, TR; DRK Photo/Tom Bean, BR. **68** Tony Waltham/Geophotos. **69** Auscape/Ferrero-Labat, TL; Auscape/Brett Gregory, TR; Martin Woodward, BL; DRK Photo/Tom Bean, BR. **70** Science Photo Library/Sinclair Stammers, BL, BM. **70–71** Science Photo Library/Vaughan Fleming. **71** Matthew White, TL; Roy Williams, ML. **72** Science Photo Library/Prof. Walter Alvarez, ML; Bradbury and Williams, B. **73** Matthew White, T; NASA/URL/(http://earthobservatory.nasa.gov), BL. **74** Matthew White, B. **74–75** Science Photo Library/Milton Heiberg. **75** Roy Williams, TL; Still Pictures/Somboon-Unep, BM. **76** Science Photo Library/Sinclair Stammers, background; Michael Holford, BM. **77** View/Paul Raftery, background; John Meek, ML. **78** Photography Norman Brand. **78–79** Crown Copyright: Historic Royal Palaces. **79** Martin Woodward, TR; W&H Synia/Covent Garden Dental Practice/photography John Meek, BR. **80** Still Pictures/Peter Frischmuth, TM. **80–81** Still Pictures/Hubert Klein. **81** Still Pictures/Roland Seitre, TR; Roy Williams, MR. **82** DRK Photo/Jeremy Woodhouse, ML. **82–83** Bradbury and Williams. **83** Matthew White, T; DRK Photo/Larry Ulrich, BR. **84** Bradbury and Williams. **84–85** Auscape/François Gohier. **85** Matthew White, MR; Science Photo Library/Geospace, BR. **86** © Royal Geographical Society, BL; Bradbury and Williams, TM. **86–87** John Cleare/Mountain Camera. **87** Bridgman Art Library/IZZA/Private Collection, TL; Mountain Camera, TR; Science Photo Library/Masao Hayashi/Duno, MR; DRK Photo/Wayne Lankinen, BR. **88** Science Photo Library/Bernhard Edmaier, TR; Still Pictures/Gil Moti, BL. **88–89** Auscape/Jean–Paul Ferrero. **89** Bradbury and Williams, TR. **90** Bradbury and Williams, ML. **90–91** DRK Photo/Michael Fogden. **91** Martin Woodward, TL; Matthew White, TR; DRK Photo/Michael Fogden, BM. **92–93** Science Photo Library/Bernhard Edmaier. **93** Science Photo Library/B&C Alexander, TR; Matthew White, MR. **94** Science Photo Library/David Nunuk, MR; Martin Woodward, MM; DRK Photo/Tom Bean, ML. **94–95** Science Photo Library/Martin Bond. **95** Still Pictures/Kevin Schafer, TR. **96** Bradbury and Williams. **96–97** DRK Photo/Stephen J. Krasemann. **97** DRK Photo/John Eastcott, MR. **98** Science Photo Library/Earth Satellite Corporation, ML; Science Photo Library/NASA, MR. **98–99** DRK/Martin Harvey. **99** DRK/David Woodfall, TR; Martin Woodward, MR. **100–101** DRK Photo/James P. Rowan. **101** DRK/David Woodfall, TL. **102** Solarfilma/GSF, MR; Matthew White, BL. **102–103** Digital Vision. **103** Matthew White, BL; DRK Photo/Norbert Wu, MR. **104–105** DRK/Fred Bruemmer. **105** Science Photo Library/CNES, MR. **106** Science Photo Library/Bernhard Edmaier, BL. **106–107** DRK Photo/Jeff Drewitz. **107** DRK Photo/Pete Oxford, TL; Auscape/Wayne Lawler, TR; DRK Photo/Larry Ulrich, MR; Roy Williams, BR. **108** Matthew White, TR; Bradbury and Williams, MM, BL; DRK Photo/Steve Kaufman, BL. **108–109** Roy Williams. **109** Bradbury and Williams, T; Image provided by ORBIMAGE. © Orbital Imaging Corporation and processing by NASA Goddard Space Flight Centre, BR. **110–111** Bradbury and Williams. **111** Bradbury and Williams/Mountain High, TM, BM. **112–113** Bradbury and Williams/Mountain High, ML; Martin Woodward, B; Bradbury and Williams, MR. **112–113** Science Photo Library/B. Murton/Southampton Oceanography Centre, background. **113** Bradbury and Williams. **114–115** Matthew White, artworks. **114–115** DRK Photo/Steve Kaufman. **115** Science Photo Library/Keith Kent, BR. **116** Bradbury and Williams. **116–117** Science Photo Library/Soames Summerhayes. **117** DRK Photo/C.C. Lockwood, MR; Science Photo Library/Bernhard Edmaier, BM. **118** Bradbury and Williams, T; Matthew White, B. **119** DRK Photo/Doug Milner. **120** Popperfoto/Reuters/Srdjan Zivulovic, BL. **120–121** Martin Woodward, artworks. **121** Popperfoto/ARC/Jean-Bernhard Sieber. **122** Still Pictures/Shehzad Noorani, BM. **122–123** Katz Pictures/Haviv/Saba-Rea. **123** Matthew White, BL. **124** Bradbury and Williams, MM. **124–125** DRK Photo/Stephen G. Maka, B. **125** Colin Woodman, MM. **126–127** Bradbury and Williams. **128** Digital Vision, TR; DRK Photo/Darrell Gulin, ML; DRK Photo/Tom Bean, BL. **128–129** Bradbury and Williams, background. **129** DRK Photo/Stephen J. Krasemann, TL; DRK Photo/Michael Fogden, MR; Science Photo Library/David Parker, BL. **130–131** Matthew White. **132** Bradbury and Williams, MM; Martin Woodward, BL. **132–133** Image provided by ORBIMAGE. © Orbital Imaging Corporation and processing by NASA Goddard Space Flight Centre. **133** Bradbury and Williams/Mountain High, TM; Science Photo Library/Brian Brake, TR. **134** Matthew White, ML; DRK Photo/John Eastcott, MR; DRK Photo/Jeremy Woodhouse, BL. **134–135** Science Photo Library/J.G. Golden. **136** John Meek, TR; DRK Photo/Tom Bean, MM; Digital Vision, BL, BR; Science Photo Library/Earth Satellite Corporation, BM. **137** DRK Photo/Stephen J. Krasemann, TL; Science Photo Library/F.S. Westmorland, MM; Digital Vision, BL; Science Photo Library/Jim Gipe, BR. **138** DRK Photo Library/Marc Epstein, ML. **139** © Crown Copyright, Met Office. Reproduced under Licence Number: Met O/IPR/2/2002/0008, TR; Steve Jebson; NERC Satellite Station, University of Dundee. **140** DRK Photo/Larry Lipsky, BL; **140–141** DRK Photo/Dick Canby. **141** NOAA, BL; Bradbury and Williams, BR. **142** Matthew White, ML; Science Photo Library/Magrath/Folsom, B. **143** top to bottom: DRK Photo/Dick Canby, 1; DRK Photo/Stephen J. Krasemann, 2; DRK Photo/M.C. Chamberlain, 3; John Meek, 4; DRK Photo/S. Nielsen, 5; Roy Williams, 6, 7. **144** Bradbury and Williams, TM; Still Pictures/Gil Moti, BM. **145** Science Photo Library/Bernhard Edmaier, background. **145** DRK Photo/D. Cavagnaro, TL; DRK Photo/Don & Pat Valenti, MR; Kevin Jones Associates, BR. **146–147** DRK Photo/Darrell Gulin. **147** DRK Photo/Wayne Lynch, TR; DRK Photo/Johnny Johnson, BL; Bradbury and Williams, BR. **148** Bradbury and Williams, MM; Photos top to bottom: DRK Photo Library/Steve Kaufman,1; Digital Vision, 2,3,4; Michael Robinson, BR. **148–149** Science Photo Library/Conor Caffrey. **149** Science Photo Library/NASA/Goddard Space Flight, TL; DRK Photo/David Woodfall, TR. **150** DRK Photo/Michael Collier, B. **150–151** Science Photo Library/Dr Morley Read. **151** Science Photo Library/Simon Fraser,TR; © Greenpeace/Rodrigo Baleia, BL. **152–153** DRK Photo/Marc Epstein.

Planet Earth and the Universe
was published by The Reader's
Digest Association Ltd, London.
It was created and produced for
Reader's Digest by Toucan Books
Ltd, London.

Questions set by
Justin Scroggie

Answers section written by
Julia Bruce, Helen Douglas-Cooper,
Justin Scroggie

For Toucan Books:
Editors
Julia Bruce, Jane Chapman, Helen
Douglas-Cooper, Daniel Gilpin,
Andrew Kerr-Jarrett
Picture researcher
Christine Vincent
Consultant
Colin Uttley
Proofreader
Ken Vickery
Indexer
Laura Hicks
Design
Bradbury and Williams

For Reader's Digest:
Project editor
Christine Noble
Project art editor
Jane McKenna
Pre-press accounts manager
Penny Grose
Editorial assistant
Katharine Swire

Reader's Digest, General Books:
Editorial director
Cortina Butler
Art director
Nick Clark

Colour origination
Colour Systems Ltd, London

Printed and bound
in Europe by Arvato, Iberia

ISBN 0 276 42715 7
BOOK CODE 625-002-02
CONCEPT CODE UK0095/G/S

First edition Copyright © 2002

The Reader's Digest Association Ltd,
11 Westferry Circus,
Canary Wharf,
London E14 4HE
www.readersdigest.co.uk

Reprinted with amendments 2003

We are committed to both the
quality of our products and
the service we provide to our
customers. We value your
comments, so please feel free
to contact us on 08705 113366
or via our web site at
www.readersdigest.co.uk

If you have any comments or
suggestions about the content
of our books, you can email us at
gbeditorial@readersdigest.co.uk

Copyright © 2002 Reader's Digest
Association Far East Limited.
Philippines copyright © 2002
Reader's Digest Association
Far East Limited